T0119346

SUPERSPREADERS OF MALIGN AND SUBVERSIVE INFORMATION ON COVID-19

Russian and Chinese Efforts Targeting the United States

MIRIAM MATTHEWS | KATYA MIGACHEVA | RYAN ANDREW BROWN

RAND
CORPORATION

For more information on this publication, visit www.rand.org/t/RRA112-11

Library of Congress Cataloging-in-Publication Data is available for this publication.
ISBN: 978-1-9774-0687-3

Published by the RAND Corporation, Santa Monica, Calif.
© 2021 RAND Corporation
RAND® is a registered trademark.

Cover: Spasskaya Tower: Сергей Шармаков/Getty Images/iStockphoto;
China national conference hall: Spondylolithesis/Getty Images/iStockphoto;
COVID-19: BlackJack3D/Getty Images/iStockphoto.
Cover design by Rick Penn-Kraus

Support RAND
Make a tax-deductible charitable contribution at
www.rand.org/giving/contribute

www.rand.org

Preface

The global spread of coronavirus disease 2019 (COVID-19) created a fertile ground for attempts to influence and destabilize different populations and countries. Both Russia and China appear to have employed information manipulation during the COVID-19 pandemic in service to their respective global agendas. This report uses exploratory qualitative analysis to systematically describe the types of COVID-19-related malign and subversive information efforts with which Russia- and China-associated outlets appear to have targeted U.S. audiences from January 2020 to July 2020 and organizes them into a framework. This work lays the foundation for a better understanding of how and whether Russia and China might act and coordinate in the domain of malign and subversive information efforts in the future. This is the first in a series of two reports, the second of which will use big data, computational linguistics, and machine learning to test findings and hypotheses generated by this report.

This report is part of RAND's Countering Truth Decay initiative, which considers the diminishing role of facts and analysis in political and civil discourse and the policymaking process. Disinformation and its rampant spread online and offline is one of the key drivers of Truth Decay. As the volume of disinformation increases and as disinformation spreads faster and further, it can create uncertainty, distrust, and confusion as it drowns out factual and objective information. Agents—notably such foreign actors as Russia and China and their proxies—fuel and contribute to the explosion in disinformation that we have observed over the past decade. Better understanding how

Russia and China operate in this space can help inform our understanding of the Truth Decay phenomenon and inform efforts to mitigate it.

Funding

Funding for this research was provided by unrestricted gifts from RAND supporters and income from operations.

Contents

Figures and Tables

Figures

Tables

Summary

Both Russia and China appear to have employed information manipulation during the coronavirus disease 2019 (COVID-19) pandemic in service of their respective global agendas. This report describes the types of COVID-19-related malign and subversive information efforts with which sources linked to Russia and to China ostensibly targeted U.S. audiences between January 2020 and July 2020. We sought two types of sources: those formally linked to Russia and China and those shown to have indirect links to Russian or Chinese governments or networks. We used exploratory qualitative analysis to describe several trends:

- Both countries disseminated messages through a wide variety of channels and platforms, including social media.
- Both countries attempted to tarnish the reputation of the United States by emphasizing challenges with its pandemic response and characterizing U.S. systems as inadequate.
- Both countries falsely accused the United States of developing and intentionally spreading the virus.
- The two countries also appeared to differ in their principal goals for COVID-19-related information efforts: Russia aimed to destabilize the United States; China aimed to protect and enhance its international reputation.
- Both countries modified their COVID-19-related messaging over time, focusing on conspiracy theories about the virus's origins and impacts from March 2020 to April 2020 and later moving to concentrate on perceived U.S. failure in responding to COVID-19.

- While Russia deployed wide-ranging media and targeted a variety of audiences, China-linked messaging was ideologically uniform, aligned across multiple information outlets, and appeared to target audiences that were less varied.

In a follow-on report, we will use big data, computational linguistics, machine learning, and other tools to examine these trends.

Implications

We further consider how characteristics of the two countries' malign and subversive information efforts could shape their future activities. Future research may help inform how best to counter such activities. We reached the following conclusions:

- Countering apparent Russian and Chinese malign and subversive information efforts will require messaging campaigns that address the capabilities and thematic emphasis of each of these actors. Future research can inform these campaigns.
- Future work that further considers which U.S. audiences are most likely to be exposed to, and influenced by, efforts from Russia and China could assist with tailoring countermessaging efforts.
- Profiling Russian and Chinese sources known to frequently create and disseminate disinformation and propaganda can also inform countermessaging efforts.
- China and Russia appear to amplify each other's messages, when opportune. This might eventually lead to some collaboration, albeit limited in nature.
- Public health messaging should account for potential impacts of Russian and Chinese messaging on vaccination uptake.

Acknowledgments

The authors would like to thank those who helped to ensure completion of this work. First, we thank Jennifer Kavanagh and Michael Rich, who provided helpful feedback regarding this research, and we thank Rich Girven, who provided thoughtful comments on the application of this research. This work has also benefited from the valuable feedback of two reviewers: Christopher Paul and Thomas Kent. We would also like to thank Erik Van Hegewald for his assistance in completing the edited report.

Abbreviations

CCP	Chinese Communist Party
COVID-19	coronavirus disease 2019
CGTN	China Global Television Network
CCTV	China Central Television
ESOC	Empirical Studies of Conflict

CHAPTER ONE

Introduction

The global spread of coronavirus disease 2019 (COVID-19) created a fertile ground for attempts to influence and destabilize different populations and countries. Russia and its agents appear to have used the COVID-19 pandemic as an opportunity to promote Russia's agenda, employing information manipulation in service to the country's apparent domestic and international objectives. China and its agents also appear to have disseminated misleading or false information involving COVID-19 to further that country's own agenda. Previous work has documented Russian and Chinese efforts to use domestic and international information spaces to achieving each country's respective goals,[1] but the pandemic offers a unique opportunity to examine how these two countries capitalized on a rapidly evolving and unstable situation.

This report describes the types of COVID-19-related malign and subversive information efforts that sources associated with Russia and China appear to have used to target U.S. audiences from January 2020 to July 2020. This report is an exploratory qualitative effort that will be followed by a later one featuring quantitative analysis using computa-

---

[1] See, for example, Miriam Matthews, Alyssa Demus, Elina Treyger, Marek N. Posard, Hilary Reininger, and Christopher Paul, *Understanding and Defending Against Russia's Malign and Subversive Information Efforts in Europe*, Santa Monica, Calif.: RAND Corporation, RR-3160-EUCOM, forthcoming; Michael J. Mazarr, Abigail Casey, Alyssa Demus, Scott W. Harold, Luke J. Matthews, Nathan Beauchamp-Mustafaga, and James Sladden, *Hostile Social Manipulation: Present Realities and Emerging Trends*, Santa Monica, Calif.: RAND Corporation, RR-2713-OSD, 2019; and Christopher Paul and Miriam Mathews, *The Russian "Firehose of Falsehood" Propaganda Model: Why It Might Work and Options to Counter It*, Santa Monica, Calif.: RAND Corporation, PE-198-OSD, 2016.

tional linguistics, machine learning, and other analytic techniques on bulk downloads of news and other data sources.

To identify messages for this qualitative effort, we searched existing databases that systematically documented Russian or Chinese efforts,[2] and we also searched for messaging efforts on English-language media channels linked to Russia and China (see the appendix). In doing so, we sought two types of sources: those formally linked to Russia and China (such as RT and CGTN, respectively) and those with indirect links to the Russian or Chinese government or Russian or Chinese networks (such as SouthFront or Global Research, respectively). We also reviewed secondary sources—such as scholarly analyses, government documents, and research reports—to gather additional information about the COVID-19-related messages promoted by Russia and China to U.S. audiences. In addition, we reviewed existing analyses of the instruments that these countries employed in spreading malign or subversive information.

Importantly, each data source has its own strengths and weaknesses, and our searches of multiple sources likely resulted in both incomplete capture and some duplicated observations. As our qualitative observations will be subjected to quantitative testing in follow-on research, we considered such sources of error as allowable for the exploratory purposes of this report.

After examining the factual accuracy of messages, we identified instances in which messages could be construed as disinformation or propaganda and analyzed the major themes. We determined whether these messages might be used on U.S. audiences by examining whether the messages were accessible within the United States and the content was delivered in English. We then developed a conceptual framework to help delineate the likely strategic goals and objectives that Russia, China, and their agents appear to have pursued in their messaging to U.S. audiences.

---

[2] "Coronavirus: Stay up to Date," EUvsDiSiNFO database, undated; Jacob N. Shapiro, Jan Oledan, and Samikshya Siwakoti, *ESOC COVID019 Disinformation Tracking Report*, Princeton, N.J.: Empirical Studies of Conflict, Princeton University, 2020; Alliance for Securing Democracy, "Hamilton 2.0 Dashboard," German Marshall Fund of the United States, undated.

To facilitate our analysis, we organized our findings by the country associated with message dissemination (either Russia or China) and considered the content and timing of messages. We then analyzed this content for its apparent purpose and possible audiences. Next, we examined the ways in which malign and subversive information efforts evolved over time and the primary instruments used for information dissemination. We then relied on these different analytic pieces to examine the possible objectives to which Russian and Chinese malign and subversive information efforts involving COVID-19 were likely contributing. Finally, we compared the approaches of both countries, identified unique and overlapping themes, and enumerated the likely synergies and differences between Russia's and China's approaches to malign information efforts and how those could affect future subversive efforts targeting the United States.

Categories of Efforts

Malign and subversive information efforts are sustained and multidimensional, involve the distribution of false or manipulated information, and appear to be conducted with political intention to shape perceptions or affect behavior. These efforts can be divided into separate categories of messaging, such as disinformation and propaganda. However, clear delineation of content into these categories is challenging because of both the fluid nature of an evolving situation (i.e., determining "truth" about the event itself) and the difficulty of inferring the precise intent behind a piece of information. In our analysis, we used specified definitions as a guide for designating pieces of information. Drawing from previous research, we defined *disinformation* as the intentional creation and/or dissemination of false information.[3] Conspiracy theories are a prominent type of disinformation involving attempts to explain the ultimate cause of events with false claims of

---

[3] Jennifer Kavanagh and Michael D. Rich, *Truth Decay: An Initial Exploration of the Diminishing Role of Facts and Analysis in American Public Life*, Santa Monica, Calif.: RAND Corporation, RR-2314-RC, 2018. For more details on our methodology, see the appendix.

secret plots by two or more powerful actors.[4] In addition, we defined *propaganda* as the intentional and systematic promotion of ideas and information that might be true (or partially true), but support a particular point of view aimed at instilling a certain attitude or response in the target audiences.[5] See Table 1.1 for the summary of these definitions.

Table 1.1
Summary of Key Definitions

Category	Definition
Malign and/ or subversive information	Activities in and through the information environment that (1) are sustained and multidimensional, (2) involve distribution of false information or information that is manipulated in other clearly identifiable ways, and (3) appear to be conducted with political intention to shape perceptions or affect behavior
Disinformation	Intentional creation and/or dissemination of false information; prominent examples are fabrication of facts, events, and conspiracy theories
Conspiracy theories	A prominent example of disinformation; attempts to explain the ultimate causes of significant social and political events (such as the pandemic) and circumstances with claims of secret plots by two or more powerful actors; can accuse any group perceived as powerful or malevolent
Propaganda	Intentional and systematic promotion of ideas and information that might be true but support a particular point of view aimed at instilling a certain attitude or response; selective use of information for political effect

SOURCES: Author's compilation of definitions from *Propaganda, Misinformation, Disinformation & Fact Finding Resources*, Detroit, Mich.: Wayne State University Library System, Research Guides, April 29, 2020; Douglas et al., 2019, p. 4; and Jackson, 2017. Also see Matthews et al., forthcoming.

---

[4] See Karen M. Douglas, Joseph E. Uscinski, Robbie M. Sutton, Aleksandra Cichocka, Turkay Nefes, Chee Siang Ang, and Farzin Deravi, "Understanding Conspiracy Theories," *Political Psychology*, Vol. 40, No. 1, 2019, p. 4.

[5] Dean Jackson, "Distinguishing Disinformation from Propaganda, Misinformation, and 'Fake News,'" Washington, D.C.: National Endowment for Democracy, Issue Brief, October 17, 2017. There is rich debate over how to conceptualize and define propaganda, and we use this working definition solely for application in this report. For additional information on the definitional debate, see Alexander V. Laskin, "Defining Propaganda: A Psychoanalytic Perspective," *Communication and the Public*, Vol. 4, No. 4, 2019; and Viorel Tutui, "Some Reflections Concerning the Problem of Defining Propaganda," *Argumentum*, Vol. 15, No. 2, 2017.

Discussions about the information space during COVID-19 also mention *misinformation* as a separate category of information that is defined as false, inaccurate, or unsupported but spread unintentionally.[6] Sources linked both to China and Russia have likely spread misinformation, but we focus in this report only on the false content that appears to be intentionally disseminated. Intention is difficult to determine, however, and the lines between disinformation and misinformation can be blurred. To err on the side of caution, we only labeled cases as disinformation when there was evidence indicating that the information was likely purposefully fabricated or spread by Chinese or Russian sources and proxies. We also considered cases to be disinformation when outlets linked to China or Russia disseminated unsupported claims in clear support of apparent strategic interests.

In the rest of this report, we will use such terms as *apparent malign and subversive information efforts* interchangeably with such shorthand terms as *information efforts* or *messaging*. We recognize that there are different degrees of linking between media sources and the official Russian and Chinese governments—from governments' direct agents to remote proxies and supporters. For simplicity, we will refer to such sources as *Russian/Chinese* or as *Russia-linked* or *China-linked*. In Chapters Two and Three, we discuss findings for Russia and China, respectively. In Chapter Four, we turn to a comparative analysis of the messaging efforts linked to these countries. We conclude this report with further discussion and implications in Chapter Five. We outline our methods in the appendix.

---

[6] *Propaganda, Misinformation, Disinformation . . .* , 2020.

CHAPTER TWO

Russia's Malign and Subversive Information Efforts

As mentioned in Chapter One, Russia and its agents appear to have used messages about COVID-19 in malign and subversive information efforts targeted to U.S. audiences. Table 2.1 presents a summary of prominent COVID-19-related content disseminated by Russia and its agents.

Disinformation

Most disinformation that appeared to be coming from Russia-linked sources can be classified as conspiracy theories. Roughly, these conspiracy theories can be divided into messages about the origins of the virus, the degree of its spread and danger, and treatments and countermeasures.

Russia-linked sources repeatedly shared conspiracy theories about the origins of the virus (see Table 2.1). For example, one conspiracy theory focused on the "world elite," claiming that the world's richest individuals and corporations instigated and funded the COVID-19 pandemic to facilitate their objectives for world domination. Another set of conspiracy theories focused on the United States as the origin of the virus; relatedly, a third set proposed that the United States purposefully manufactured the virus in one of its domestic or international labs with the aim of weakening China or Russia. Although there was variability in the specific content of these conspiracy theories, the common thread proposing that the United States had developed the virus as a means for sustaining its global domination and curbing the assent of its adversaries is evident.

7

Table 2.1
Examples of COVID-19 Disinformation Content Associated with Russia and Its Agents

Topic	Example Content	Example Source
Conspiracy theories on the origin of the virus	• To control and/or reduce world populations, Bill Gates and other world elites masterminded and funded the pandemic, then forced global organizations to support them in spreading panic.	Peter Koenig, "The Coronavirus COVID-19 Pandemic: The Real Danger is Agenda ID2020," GlobalResearch, March 12, 2020
	• COVID-19 was developed by the United States to weaken China or Russia and sustain global domination. • U.S. Secretary of State Mike Pompeo accidentally admitted that COVID-19 is a live military exercise. • COVID-19 is likely a genetically engineered "supergerm" designed to let those with money take over the world.	Phil Butler, "Is COVID-19 Devouring What's Left of the Trump Presidency?" *New Eastern Outlook*, March 16, 2020
Conspiracy theories on the spread and danger of the virus	• The spread and lethality of the virus have been exaggerated to impose greater control over the population and to deprive Americans of their constitutional rights. • The United States has become a socialist country that severely curtails civil liberties. • U.S. residents will be tracked by their health records and vaccination statuses and will be denied service (e.g., by an ATM) if unvaccinated. • The response to the pandemic is a successful experiment in global manipulation and population subjugation.	"COVID-19 Eroding Civil Liberties—Gone for Good?" RT America, via YouTube, March 31, 2020 Slobodan Solajic, "The COVID-19 Plandemic Is an Experiment in Manipulating the World," One World Global Think Tank, June 16, 2020

Table 2.1—Continued

Topic	Example Content	Example Source
Conspiracy theories on treatment of and countermeasures for the virus	• The United States is preventing affordable COVID-19 treatments from reaching markets to ensure profits for its pharmaceutical companies. • Everything is profit-driven in the West. With different vaccines from different pharmaceutical giants (known as Big Pharma) coming on the market, who will tell the patient which one is the best or most suitable for the patient's condition? This will lead to a chaotic scam on citizens. • Nevada's medical team, which banned COVID-19 treatments using federally approved malaria drugs chloroquine and hydroxychloroquine, is unprofessional and unqualified. • The Western media, in collaboration with Big Pharma, ignores the fact that high doses of vitamin C saved lives in China. Reports of this cheap, safe treatment pioneered in China have been completely ignored by the West because leaders are beholden to the Big Pharma approach to the current pandemic.	"Is There a Doctor in the House? Nevada State Medical Team That Banned Malaria Drug for Coronavirus Lacks Qualification," RT, March 25, 2020
	• The World Health Organization, which has links to the United States through Bill Gates, tried to silence Sweden's effective approach.	Martin Armstrong and Joaquin Flores, "World Health Organization Demanded Sweden Lock Down to Cover Its Own Fraud," Fort Russ News, April 17, 2020
U.S. global action	• The United States uses sanctions in Iran, Syria, Venezuela, and elsewhere during the COVID-19 crisis as a tool to overturn the regimes in these countries.	Andrew Korybko, "The US' Medical Terrorism Against Syria Threatens to Make World War C Much Deadlier," One World Global Think Tank, May 20, 2020

Table 2.1—Continued

Topic	Example Content	Example Source
Other messages on spread and danger of the virus	• It is too early to accurately gauge how many—if any—extra people will die because of COVID-19. • It is hard to believe that when this all blows over, the damage that will have been done by the shutdown measures—to businesses, civil liberties, individual lives and the global economy—could have been for nothing. Nonetheless, it seems entirely possible that this will be the case.	Peter Andrews, "COVID-19's Meant to Be a New Black Death, but in Britain No More People Are Dying Than NORMAL. What Does This Say About the Virus?" RT, March 31, 2020
Other messages on treatment and pandemic response	• Many other countries—e.g., Mexico, Bolivia, Russia, China—have been doing a much better job than the United States in responding to the pandemic.	"America's China Blame Game Masks COVID-19 Failures (full show)," RT America, via YouTube, May 1, 2020
	• Despite spending billions of dollars on technology to detect pandemic outbreaks at an early stage, the United States still was not prepared to handle the COVID-19 crisis.	"How the US Overlooked the COVID Outbreak (RT documentary)," RT America, via YouTube, April 18, 2020
	• China's successful handling of the pandemic stems from its understanding of human rights, unlike the United States and Europe.	John Ross, "Human Life Must Trump Economics in a Pandemic. THIS Is Why China Is Succeeding in War on COVID-19 and US Is on Path to Disaster," RT, March 30, 2020

Russia-linked sources also promoted disinformation about the spread of the virus and the available treatments. To this end, conspiracy theories promoted an idea that the wide spread of the virus was greatly exaggerated to benefit U.S. corporations. Other related theories suggested that the U.S. government inflated numbers regarding the spread to scare Americans into giving up their constitutional rights (purportedly an irreversible process).

Focusing on messages about treatments and countermeasures, Russia-linked sources peddled conspiracy theories that claimed the suppression of evidence for effective and easily accessible treatments ranging from hydroxychloroquine to vitamins and herbal tinctures. Some of these theories proposed that the reason for this alleged suppression were that profit-seeking pharmaceutical companies prioritized expensive developments of new medical treatments and vaccinations and that "elites" were seeking world domination through control of medicines or through their use. Other conspiracy theories said these same groups were planning to use vaccination as instruments for population monitoring and control.

Beyond conspiracy theories, other disinformation messages connected to Russia or its agents also perpetrated false or unconfirmed information (see Table 2.1). Among such messages are discussions of the likely overestimation of the danger of COVID-19 and the unnecessarily restrictive measures to stop its spread. It is plausible that these comparisons were a product of attempts to find meaning in an evolving situation rather than of intentional deceit, but such messages often appeared in concert with suggestions that the U.S. government or its "elites" were exaggerating the virus' spread to control the U.S. population—a claim that more clearly falls into the disinformation category. We also categorized as disinformation the claims that the United States maintained sanctions on Iran, Syria, and Venezuela as a way of using the crisis to overturn the ruling regimes in these countries.

Propaganda

In addition to conspiracy theories, a large proportion of messaging that appeared connected to Russia or its agents focused on spreading and amplifying negative information about the United States; we classified such content as *propaganda*. Although some of these reports contained some truthful content, they tended to focus disproportionately on negative COVID-19-related developments in the United States and contributed to a broader context of negatively slanted reporting about the United States. The U.S. response to the virus was the most prominent focus of the messages in this category; these messages highlighted supposed inefficiencies within the U.S. medical system, apparent inadequacy of measures to curb the pandemic within the country, and potential economic implications of, or reasons for, the U.S. response. Frequently, these messages unfavorably compared the U.S. "failure to adequately respond" with the success of other countries, most often China. Earlier in the pandemic (until the first prominent wave of cases hit Russia in April 2020), Russia-linked sources also presented Russia as an example of a public health response superior to that of the United States. Next to Russia's and China's more-authoritarian governments, the messages connected to Russia or its agents implied or directly proposed that the more-democratic U.S. systems were poorly equipped to deal with a pandemic-level crisis.

Trends in Content over Time

Russia-linked messaging to U.S. audiences that was related to COVID-19 appeared to change focus and priorities over time (see Figure 2.1). Earlier in 2020, particularly between March and April, Russia-linked messaging seemed to place greater emphasis on conspir-

Figure 2.1
Trends in Content on COVID-19 from Russia and Its Agents in the First Half of 2020

NOTE: Darker shades of blue indicate greater intensity of messaging at different points during the time frame that we analyzed.

acy theories that focused on the origins and spread of the virus and on supposed forced vaccination. As the pandemic evolved, the messaging shifted in mid-April and early May 2020 toward emphasizing U.S. flaws in responding to the crisis. Russian messaging also appeared to place an increased emphasis on contrasting the U.S. COVID-19 response with that of China. Portraying the United States as unfairly attacking China and instigating an information war against China also became increasingly prominent in Russia-linked messaging. The volume of COVID-19 messaging from Russia and its agents notably waned toward late May and early June 2020 as the focus of the content appeared to shift to coverage of racial justice protests and related controversies in the United States.

Instruments

Several Russia-linked outlets helped propagate the malign and subversive COVID-19 information efforts that targeted U.S. audiences. Some of these outlets have direct, open relationships with the Russian government—such as RT America, Sputnik, and their associated Facebook and YouTube channels.[1] Others are media sources with ties to the Russian government through its associates—such as SouthFront, News Front, New Eastern Outlook Journal, Global Research, Katehon, the Strategic Culture Foundation, and One World Press (see Table 2.2 for brief descriptions of each source).[2] Together, these outlets represent different ideologies and objectives; thus, they attract a wide variety of audiences. RT and Sputnik tended to engage in propaganda; other outlets, such as SouthFront and Global Research, focused more heavily on disinformation and conspiracy theories.[3]

---

[1] RT also has newer subsidiaries, such as Redfish, a platform that produces political videos and appeals to younger audiences.

[2] "The Community of Collapse," EUvsDiSiNFO, March 27, 2020.

[3] Global Engagement Center, *GEC Special Report: Pillars of Russia's Disinformation and Propaganda Ecosystem*, Washington, D.C.: U.S. Department of State, August 2020.

Table 2.2
Description of Russia-Linked Disinformation Sources

Source	Description	Primary audience
RT America	• U.S.-based pay television and internet-based news channel • Part of the global multilingual RT network • Funded and controlled by the Russian government	Variety of audiences with larger appeal to those closer to the margins of the U.S. political spectrum
Sputnik	• English-language news website platform and radio broadcast service • Russian state-owned news agency, headquartered in Moscow	U.S. coverage leans to the right politically
Redfish	• New, left-leaning subsidiary of Ruptly (which, in turn, is a subsidiary of RT)	Politically left-leaning, youth
SouthFront; Analysis and Intelligence (a.k.a. SouthFront)	• A multilingual online disinformation site registered in Russia with evidence of connection to News Front and other Russian sources; hides connections to Russia • Focuses on military and security issues; combines Kremlin talking points with detailed knowledge of military systems and ongoing conflicts; uses flashy maps, infographics, and videos	Aims to appeal to military enthusiasts, veterans, and conspiracy theorists
News Front	• Crimea-based outlet with self-proclaimed goal of providing an "alternative source of information" for Western audiences	Nationalist, right-leaning
New Eastern Outlook Journal	• English-language, pseudo-academic publication of the Russian Academy of Science's Institute of Oriental Studies • Promotes disinformation and propaganda focused primarily on the Middle East, Asia, and Africa • Combines pro-Kremlin views of Russian academics with anti-U.S. views of Western fringe voices and conspiracy theorists	Conspiracy theory enthusiasts, critics of U.S.-led global policy

Table 2.2—Continued

Source	Description	Primary audience
Global Research	• Canada-based website that has become deeply enmeshed in Russia's broader disinformation and propaganda ecosystem • Draws fringe authors and conspiracy theorists from Global Research contributors • Provides a Western voice to Russian- or Kremlin-supported narratives	Conspiracy theory enthusiasts
The Strategic Culture Foundation	• Online journal registered in Russia, directed by Russia's Foreign Intelligence Service (SVR); closely affiliated with the Russian Ministry of Foreign Affairs • Russian origins of the journal obscured • Plays a central role among a group of linked websites that proliferate Russian disinformation and propaganda • Attracts Western fringe thinkers and conspiracy theorists, giving them a broader platform	Conspiracy theory enthusiasts, those on the margins of political spectrum
Geopolitica.ru	• A platform for Russian ultra-nationalists, rooted in Dugin's Eurasianist ideology • Managed by Russian nationalist oligarch Konstantin Malofeev • Spreads disinformation and propaganda targeting Western and other audiences • Claims to fight the information war against Western ideals of democracy and liberalism; publishes in English, Russian and seven other languages	Ultra-nationalists, far-right conservatives
Katehon	• Moscow-based think tank managed by Konstantin Malofeev • Disseminates anti-Western disinformation and propaganda via a multilingual website • Claims to provide independent, analytical material aimed largely at European audiences with content dedicated to "the creation and defense of a secure, democratic and just international system"	Right, conspiracy theory enthusiasts

Table 2.2—Continued

Source	Description	Primary audience
Fort Russ News	• Pro-Kremlin talking points and apparent agenda • Claims to be a "team of freedom fighters doing journalism . . . with a focus primarily on the trials and tribulations of modernity, and anti-imperialist struggles world-wide which signal the rise of a multipolar world"[a]	Conspiracy theory enthusiasts; critics of U.S.-led global policy
One World Press	• Pro-Kremlin talking points • "A new analytical and nonprofit start-up. Our media outlet aims to connect people from all across the world who share the same interest in international events. We don't have any 'political correctness' or agenda, and we actually encourage constructive and respectfully expressed criticism of anyone and anything."[b]	Conspiracy theory enthusiasts

SOURCES: Unless otherwise indicated, descriptions in this table are drawn from Global Engagement Center, 2020. Additional sources were also consulted, such as "Media Bias/Fact Check: The Most Comprehensive Media Bias Resource," Sputnik News, webpage, undated; and "Community of Collapse," 2020.

NOTES: Assumptions about primary audiences for these sites are derived from our analysis of the sites' primary content, psychological research on how people tend to seek information that confirms their world views, and existing analytical accounts.

[a] Joaquin Flores, Tom Winter, and Drago Bosnic, eds., "About FRN," Fort Russ News, 2014.

[b] OneWorld Global Think Tank, "About Us," webpage, undated.

Notably, Russia-led disinformation efforts also utilized social media. Although our analysis did not focus on the messaging spread by trolls and bots on social media, multiple sources point to these as critical tools in dissemination and amplification of disinformation.[4] Like other outlets, fake social media accounts do not appear to align under a unifying guiding ideology; instead, they spread a variety of messages that contribute to controversies and general information disarray.[5]

---

[4]   For example, see Darren L. Linvill and Patrick Warren, "Yes, Russia Spreads Coronavirus Lies. But They Were Made in America," *Washington Post*, April 2, 2020.

[5]   For example, see Matthew Rosenberg, Nicole Perlroth, and David E. Sanger, "'Chaos Is the Point': Russian Hackers and Trolls Grow Stealthier in 2020," *New York Times*, January 10, 2020.

Audiences

The information propagated by Russia-linked sources appears to have been designed to resonate with diverse audiences in the United States. We first discuss the appeal of Russia-linked COVID-19 content regarding conspiracy theories, then turn to the broad appeal of Russia-linked COVID-19 propaganda.

Some research on the characteristics of conspiracy theory enthusiasts suggests that individuals with more-conservative sociopolitical orientations (and higher levels of right-wing authoritarianism[6]) are most likely to endorse conspiracy beliefs.[7] However, emerging evidence indicates that people on both margins of the political spectrum,[8] and

---

[6] *Right-wing authoritarianism* is defined as a political attitude characterized by obedience to an authoritarian leader, a deep embrace of traditional societal values, and, at the same time, a distrust of governmental structures. Roland Imhoff and Martin Bruder, "Speaking (Un-) Truth to Power: Conspiracy Mentality as a Generalised Political Attitude," *European Journal of Personality*, Vol. 28, No. 1, 2014; Sean Richey "A Birther and a Truther: The Influence of the Authoritarian Personality on Conspiracy Beliefs," *Politics & Policy*, Vol. 45, No. 3, 2017.

[7] Martin Bruder, Peter Haffke, Nick Neave, Nina Nouripanah, and Roland Imhoff, "Measuring Individual Differences in Generic Beliefs in Conspiracy Theories Across Cultures: Conspiracy Mentality Questionnaire," *Frontiers in Psychology*, Vol. 4, No. 225, April 30, 2013; Natasha Galliford and Adrian Furnham, "Individual Difference Factors and Beliefs in Medical and Political Conspiracy Theories," *Scandinavian Journal of Psychology*, Vol. 58, No. 5, 2017; Monika Grzesiak-Feldman and Monika Irzycka, "Right-Wing Authoritarianism and Conspiracy Thinking in a Polish Sample," *Psychological Reports*, Vol. 105, No. 2, 2009; Adam M. Enders and Steven M. Smallpage, "Informational Cues, Partisan-Motivated Reasoning, and the Manipulation of Conspiracy Beliefs," *Political Communication*, Vol. 36, No. 1, 2019; Joanne M. Miller, Kyle L. Saunders, and Christina E. Farhart, "Conspiracy Endorsement as Motivated Reasoning: The Moderating Roles of Political Knowledge and Trust," *American Journal of Political Science*, Vol. 60, No. 4, 2016.

[8] Andreas Goreis and Martin Voracek, "A Systematic Review and Meta-Analysis of Psychological Research on Conspiracy Beliefs: Field Characteristics, Measurement Instruments, and Associations with Personality Traits," *Frontiers in Psychology*, Vol. 10, No. 205, February 11, 2019; Jan-Willem van Prooijen, André P. M. Krouwel, and Thomas V. Pollet, "Political Extremism Predicts Belief in Conspiracy Theories," *Social Psychological and Personality Science*, Vol. 6, No. 5, 2015; Y. Lahrach and Adrian Furnham, "Are Modern Health Worries Associated with Medical Conspiracy Theories?" *Journal of Psychosomatic Research*, Vol. 99, 2017.

those who are independents,[9] can also endorse conspiratorial content if it aligns with their predispositions and cognitive biases.[10] Some of the general characteristics of conspiracy theory enthusiasts are strong identification with their groups (e.g., political parties), feelings that their groups are treated unfairly, entitled to more, and under threat; and a sense of powerlessness and uncertainty.[11]

The wide variety of conspiracy messages disseminated by Russia-linked media sources, the polarized environment within the United States,[12] and the uncertainty of the pandemic all created a fertile ground for potential receptivity to conspiracy messaging.[13] Recent research that investigated support for conspiracy theories regarding COVID-19 found that supporters of President Donald Trump were more likely to believe that the spread of the virus has been exaggerated to damage his chances for reelection.[14] However, that study also found that conservatives and liberals were equally likely to endorse the belief that coronavirus was created and spread intentionally. Similarly, other Russia-backed conspiracy

---

[9]   Joseph E. Uscinski and Joseph M. Parent, *American Conspiracy Theories*, New York: Oxford University Press, 2014; Joseph E. Uscinski, Casey Klofstad, and Matthew D. Atkinson, "What Drives Conspiratorial Beliefs? The Role of Informational Cues and Predispositions," *Political Research Quarterly*, Vol. 69, No. 1, 2016.

[10]   Joseph E. Uscinski, Adam M. Enders, Casey Klofstad, Michelle Seelig, John Funchion, Caleb Everett, Stefan Wuchty, Kamal Premaratne, and Manohar Murthi, "Why Do People Believe COVID-19 Conspiracy Theories?" *Misinformation Review*, Vol. 1, No. 3, April 28, 2020.

[11]   Douglas et al., 2019; John T. Jost, Jack Glaser, Arie W. Kruglanski, and Frank J. Sulloway, "Political Conservatism as Motivated Social Cognition," *Psychological Bulletin*, Vol. 129, No. 3, 2003; Agnieszka Golec de Zavala and Christopher M. Federico, "Collective Narcissism and the Growth of Conspiracy Thinking over the Course of the 2016 United States Presidential Election: A Longitudinal Analysis," *European Journal of Social Psychology*, Vol. 48, No. 7, 2018; Jan-Willem Van Prooijen and Michele Acker, "The Influence of Control on Belief in Conspiracy Theories: Conceptual and Applied Extensions," *Applied Cognitive Psychology*, Vol. 29, No. 5, 2015.

[12]   Michael Dimock and Richard Wike, "America Is Exceptional in the Nature of Its Political Divide," *FactTank News*, Pew Research Center, November 13, 2020.

[13]   Hannah Rettie and Jo Daniels, "Coping and Tolerance of Uncertainty: Predictors and Mediators of Mental Health During the COVID-19 Pandemic," *American Psychologist*, August 2020.

[14]   Uscinski et al., 2020.

theories might hold appeal across the political spectrum. For example, those on the left might be particularly vulnerable to conspiracy theories about the suppression of cheap, effective treatments for the sake of pharmaceutical corporation profits. Meanwhile, conspiracy theories about forced vaccinations might appeal to libertarians, conservatives on the right, and—to a slightly lesser extent—far-left liberals who do not trust government medical experts.[15] Messages on masks and social-distancing rules as forms of population control and constitutional impingement might appeal to government and science skeptics more generally.

Russia-linked propaganda also seems to have been designed to appeal to audiences across the political spectrum. For example, messages about U.S. failures to effectively counter the spread of the virus could have appealed to the moderate left that viewed the Trump administration with skepticism. Meanwhile, messages that compared the inefficiency of U.S. systems with the better approaches in China could have appealed to those on the left of U.S. political margins. The outreach to broader audiences is also reflected in the variety of Russia-linked outlets and their ideological leanings: Some outlets seem to be designed to appeal mostly to the left (e.g., Redfish); information propagated by others (e.g., SouthFront) would mostly appeal to those on the right. The programming on RT America also seems designed to have broad appeal; the same RT opinion show might discuss the need for socialized medicine (a message in alignment with the left ideology) in

---

[15] Jennifer Reich, a sociologist at the University of Colorado Denver, who studies the spread of misinformation about health, noted that the anti-vaccination movement is "one of those places where right meets left, and that's been true for a long time" (Kiera Butler, "The Anti-Vax Movement's Radical Shift from Crunchy Granola Purists to Far-Right Crusaders," *Mother Jones*, June 18, 2020). People on the left might have anti-vaccine attitudes because of their search for "purity;" people on the right sought "liberty" (Avnika B. Amin, Robert A. Bednarczyk, Cara E. Ray, Kala J. Melchiori, Jesse Graham, Jeffrey R. Huntsinger, and Saad B. Omer, "Association of Moral Values with Vaccine Hesitancy," *Nature Human Behaviour*, Vol. 1, No. 12, December 2017). Also see Bert Baumgaertner, Julie E. Carlisle, and Florian Justwan, "The Influence of Political Ideology and Trust on Willingness to Vaccinate," *PloS One*, Vol. 13, No. 1, January 25, 2018; Mitchell Rabinowitz, Lauren Latella, Chadly Stern, and John T. Jost, "Beliefs About Childhood Vaccination in the United States: Political Ideology, False Consensus, And the Illusion of Uniqueness," *PloS One*, Vol. 11, No. 7, July 8, 2016.

one episode and condemn the United States as a "socialist country with all the handouts" in another.

Put together, the different types of messages propagated by Russia-linked sources could appeal to a variety of audiences across the political spectrum: Trump supporters on the moderate and far right and Trump skeptics on the moderate and far left, capitalism skeptics on farther edges of the left and capitalism supporters on the right, and conspiracy enthusiasts and those who disagree with the use of vaccines (commonly known as anti-vaxxers) on both the left and right margins of the political spectrum. See Figure 2.2 for a visual summary of U.S. audiences deemed potentially receptive to Russia-linked messaging on COVID-19.

Figure 2.2
2020 Audiences Potentially Vulnerable to Messages on COVID-19 from Russia and Its Agents and Proxies

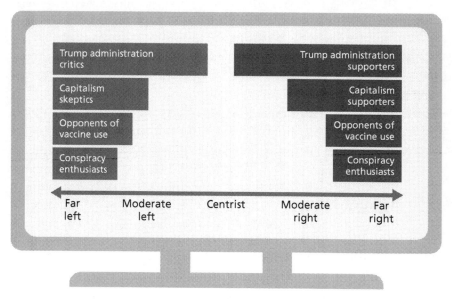

NOTES: The arrow symbolizes the U.S. political spectrum. The length of blue and red shading of audience categories reflect where on this spectrum the people within these categories could fall.

Putting it Together: Objectives

We identified four apparent strategic objectives for Russian COVID-19 messaging to U.S. audiences:

1. *Sow distrust* regarding the origin and spread of the virus and regarding countermeasures taken in the West.
2. *Amplify internal divides* by promoting negative content that appeals to people on different ends of the U.S. political and ideological spectrums.
3. *Pollute information space* by endorsing multiple versions of unconfirmed information, making it harder to differentiate the truth from the myth and contributing to the overall sense of cognitive fatigue.
4. *Discredit the U.S. image* by emphasizing its failures, especially in contrast to China's (and, sometimes Russia's) purported successes in handling the pandemic.

See Table 2.3 for examples of the alignment between the content of messages and their possible objectives. Notably, similar messages could simultaneously contribute to different objectives.

The way that COVID-19 messages from Russia developed over time suggests a broader destabilization goal underlying this content. During the pandemic, Russia-linked information efforts emphasized divisive and polarizing topics. Conspiracy theories—especially those with polarizing implications (e.g., malign plots by world elites)—were prevalent in March and early April 2020 when the situation was evolving and the United States was ramping up its pandemic response. These early efforts likely aimed to (1) capitalize on the general sense of uncertainty, fear, and apprehension that often occurs during a crisis,[16] (2) plant seeds of distrust regarding official information, and (3) test the capacity for partisan messaging. As U.S. officials engaged in debates about the proper response to the pandemic, Russian out-

---

[16] Katherine L. Einstein and David M. Glick, *Scandals, Conspiracies and the Vicious Cycle of Cynicism*, Chicago, Ill.: 2013 Annual Meeting of the American Political Science Association, working paper, August 29–September 1, 2013.

Table 2.3
Alignment Between Russian Content and Likely Objectives

Messages	Likely Objectives
• The spread and lethality of the virus have been exaggerated to exert greater control over the population and to deprive Americans of their constitutional rights.	Sow distrust
• The U.S. government is keeping the real situation with the COVID-19 pandemic a secret, even from state governors.	
• The Western media, in collaboration with Big Pharma, ignores the fact that the high doses of vitamin C saved lives in China. Reports of this cheap, safe treatment pioneered in China have been completely ignored by Western leaders because they are beholden to the Big Pharma approach to the pandemic.	
• The United States has become a socialist country that severely curtails civil liberties.	Amplify internal divides
• U.S. residents will be tracked via their health and vaccination status and will be denied service (e.g., by an ATM) if unvaccinated.	
• The pandemic test revealed the inefficiency and inhumanity of the U.S. health system.	
• COVID-19 mortality rates are overblown; it is not much worse than flu.	Pollute information space
• The goal of the elite within the World Economic Forum (e.g., the Rockefellers, Rothschilds, Morgans) is population reduction so that a small elite can benefit from the Earth's limited resources.	

lets often focused on the divides brought about or exacerbated by the pandemic response, amplifying them further. COVID-19 messaging appeared to subside in late May and early June 2020, when new potentially polarizing issues arose (e.g., racial justice protests). At that point, Russia's information efforts seemed to shift to racial divides within the United States—likely because these new issues offered better opportunity for stirring volatility.

Finally, the wide-ranging content, methods, and outlets of dissemination of information point toward the goal of appealing to a variety of audiences, with a particular focus on the margins of the U.S.

Table 2.3—Continued

Messages	Likely Objectives
• COVID-19 was developed by the United States to weaken China or Russia and to sustain global domination.	Discredit U.S. image
• The COVID-19 crisis has revealed the truth about the malignant character of the U.S. ruling class.	
• Elsewhere in the world, the crisis is going to get worse before it gets better, but it eventually will get better. In the United States, on the other hand, things will only get much worse.	
• The decadent West is committing suicide through its irrational response to COVID-19. Hatred of centuries-long values of other nations is the true threat.	
• An increasingly decadent mass culture has developed in opposition to the traditional values of Western civilization. People accept the suicidal shutting down of their economies to prevent old and sick people from dying just a few weeks or months before they would have died anyway.	
• Sacrificing economies for the sake of beating COVID-19 is akin to a medieval witch hunt; the sacrifice becomes more important than finding an effective method to deal with the problem.	
• Many Western leaders worried about their political survival delayed taking preventive actions to stem community transmission of COVID-19.	
• Unlike citizens in Western democracies (where political culture is shaped by asserting individualism, human rights, and treating the state as an enemy of liberty), Asians mostly do not worry about violation of privacy or government abuses.	
• Authoritarian societies were the only ones effective in their response to COVID-19. Democracies in the United States and elsewhere have failed in their responses, proving that neoliberalism and democracy are flawed and inferior systems; China's success is rooted in the true understanding of what human rights are—it is a right to live, first and foremost.	

political spectrum. These outlets provided COVID-19-related information fit to fill different echo chambers. Because of reliance on You-Tube and its algorithms for content dissemination, even the channels that tend to argue that they do not have a political agenda and are trying to objectively cover all points of view (e.g., RT) inevitably contribute information to echo chambers, thereby further widening existing divides.[17]

Russia Summary

Our analysis of messages that were distributed from January 2020 to July 2020 suggests that Russia engaged in apparent malign and subversive information efforts, such as disinformation and propaganda, on the topic of COVID-19. The diversity of the messages and sources could appeal to various audiences and seem designed to appeal to those across the U.S. political spectrum. Emphasizing divisive content, these information efforts appear to have sought to sow distrust within the United States, amplify existing sociopolitical divides, damage the U.S. image, and pollute information space. Achieving these objectives appears to serve the overarching strategic goal to discredit, weaken, and destabilize the United States (see Figure 2.3).

---

[17] Alessandro Bessi, Fabiana Zollo, Michela Del Vicario, Michelangelo Puliga, Antonio Scala, Guidro Caldarelli, Brian Uzzi, and Walter "Users Polarization on Facebook and YouTube," *PloS One*, Vol. 11, No. 8, August 23, 2016. Also see Jonas Kaiser and Adrian Rauchfleisch, "Unite the Right? How YouTube's Recommendation Algorithm Connects the US Far-Right," *Medium*, April 11, 2018; Roland G. Fryer, Jr., Philipp Harms, and Matthew O. Jackson, "Updating Beliefs when Evidence is Open to Interpretation: Implications for Bias and Polarization," *Journal of the European Economic Association*, Vol. 17, No. 5, October 2019.

Figure 2.3
Summary of Findings on Russia-Linked COVID-19 Messaging

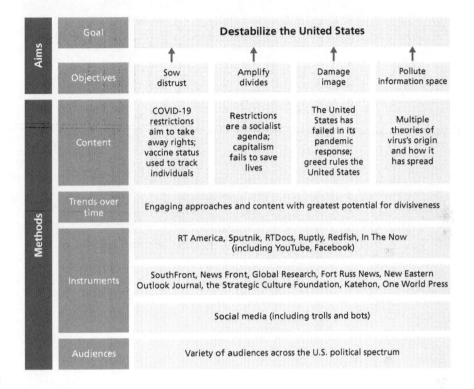

CHAPTER THREE

China's Malign and Subversive Information Efforts

China also proved to be a prolific "supplier" of disinformation and propaganda targeting U.S. audiences during the COVID-19 pandemic. Aiming to defend and protect its image, China mobilized its English-speaking television networks, internet publications, and social media tools to promote its objectives. Despite the variety of outlets and messengers, we note a generally unified, ideologically bound voice across different platforms of Chinese origin. We organized Chinese messaging efforts involving COVID-19 and targeting U.S. audiences into the categories of disinformation and propaganda. Table 3.1 presents a summary of some of the prominent messages in each category.

Disinformation

China-linked disinformation efforts focused on the origins of the virus and involved multiple conspiracy theories, such as that the virus originated in the United States or in one of its global biolabs and traveled from the United States to China and other countries (e.g., Canada; see Table 3.1). In one version, Chinese Foreign Ministry official Zhao Lijian retweeted an article from a Kremlin-linked source, Global Research, stating that the virus originated in the United States and was brought to China by the U.S. military. Chinese media also suggested that the United States was covering up the true start date of the virus spread

Table 3.1
Examples of COVID-19 Disinformation Content Associated with China and Its Agents and Proxies

Topic	Example Content	Example Source
Origins	Zhao Lijian, Chinese Foreign Ministry deputy director-general of the information department, tweeted an article from a known pro-Kremlin source claiming that coronavirus originated in the United States: "This article is very much important to each and every one of us. Please read and retweet it. COVID-19: Further Evidence That the Virus Originated in the US."	Zhao Lijian [@zlj517], Twitter post, March 12, 2020
	"Chinese Foreign Ministry spokesperson Hua Chunying said on Wednesday that the American people and the world need to know when the COVID-19 virus first started in the United States, in response to U.S. Secretary of State Mike Pompeo's recent attack on the Communist Party of China (CPC) over its response to the outbreak."	"Chinese Official: World Needs to Know . . . " April 29, 2020
	"It's getting harder to find the true origin of the novel coronavirus that has plagued the world, killing hundreds of thousands as genetic analysis indicates that the beginning of the pandemic dates back to early October 2019, over a month earlier than what was previously thought."	"COVID-19 Spread Started as Early as October 6, 2019: Study," CGTN, May 6, 2020
	Geng Shuang, a Foreign Ministry spokesperson, tweeted, "The local people are deeply concerned about the function, purpose, and safety of the biological labs in the former Soviet Union countries which were established by the United States. The United States should explain and address these concerns."	Spokesperson 发言人办公室 [@MFA_China], Twitter post, April 29, 2020
	"Why has the U.S. built so many biological laboratories in other countries? What's the purpose in locating these laboratories in countries surrounding China and Russia? Do these laboratories meet security standards? Are there hidden dangers of a leak? When can the U.S. respond to the various suspicions concerning its overseas labs?"	Le Yu, "Four Questions the US Must Answer Concerning COVID-19," *Global Times*, May 1, 2020
Spread and danger	The spread of COVID-19 is higher than previously thought, but that also means that the mortality rate is lower than previously thought.	"Trump's Anti-Coronavirus Farce," CGTN America, via YouTube, May 26, 2020

Table 3.1—Continued

Topic	Example Content	Example Source
Treatment and pandemic response	The United States is going under lockdown—gather food and all needed supplies, because soon people will not be able to leave the house for a long time.	Wong, Rosenberg, and Barnes, 2021
	Greed will guide vaccine development and other efforts in the United States; China will share with the rest of the world, but the United States will keep resources for itself.	"Commentary: A Disease-Ridden U.S. Fails World in Anti-Virus Cooperation," Xinhua, November 15, 2020; "GLOBALink | Brazilian Expert Applauds China-Brazil Cooperation in COVID-19 Vaccines," Xinhua, January 9, 2021
	As world leaders laid out the strategy, Trump was not worried. He has a talent for shielding himself from warnings from professionals, scientists, and international actors. Can Americans be reassured by his response? Trump felt the economic pain of the pandemic, not the pain of losing human lives. More Americans have died of COVID-19-related causes than died in combat during the Vietnam War.	"Trump's Anti-Coronavirus Farce," 2020
	Political ads on Facebook and Instagram praised Beijing and bashed Trump.	Hu Xijin, "[#HuSays] Trump refused to participate in the #WHA and sent US HHS Secretary Alex Azar in his place . . . ," *Global Times* post on Facebook, May 20, 2020; Hu Xijin, "The #US government's arrogance is leading the country down a road devoid of science or rationality . . . ," *Global Times* post on Facebook, August 5, 2020
	Late response to COVID-19, failure to provide personal protective equipment, failure to provide people with access to health care, and the ongoing debate about wearing a mask all indicate that the United States has failed to act appropriately.	"How the U.S. Response to COVID-19 Failed and Caused Thousands of Deaths," CGTN America, via YouTube, May 25, 2020

Table 3.1—Continued

Topic	Example Content	Example Source
U.S. global action	The United States used COVID-19 challenges as cover to incite the coup in Venezuela: "China warns it will not accept any U.S. invasion of Venezuela."	Qiao Collective [@qiaocollective], "China warns that it will not accept any US invasion of Venezuela," Twitter post, May 6, 2020

in the United States to obfuscate the truth about the virus' origins.[1] Other cases of disinformation, particularly in January and February 2020 (before the spread of the coronavirus grew globally), claimed that the United States suppressed information about COVID-19.[2]

Disinformation also touched on treatments and countermeasures. In one March 2020 case, early in the pandemic, a China-linked source spread a message via social media and text messaging apps that warned Americans of the impending lockdown by the U.S. federal government.[3] Disinformation content also featured suggestions that greed would likely drive all U.S. countermeasures to coronavirus, such as vaccines and effective treatment. Such messages purported that China would develop the vaccine for the global public good, but the United States would ensure that the rest of the world would not get the vaccine until it saturated its own needs—and that the primary goal of U.S. corporations would be to enrich themselves. Other disinformation claimed that the United States was using COVID-19 as a cover-up for foreign interference, such as an alleged attempt at sparking a coup in Venezuela.[4]

---

[1] "Chinese Official: World Needs to Know When Coronavirus Started in U.S.," CGTN, April 29, 2020.

[2] "China Delayed Releasing Coronavirus Information for Days at Start of Outbreak, Frustrating the WHO," *Los Angeles Times*, June 2, 2020.

[3] Edward Wong, Matthew Rosenberg, and Julian E. Barnes, "Chinese Agents Helped Spread Messages That Sowed Virus Panic in US, Officials Say," *New York Times*, April 22, 2020.

[4] Amber Frankland, Bret Schafer, and Matt Schrader, "Hamilton Weekly Report: May 2–8, 2020," Alliance for Securing Democracy, May 11, 2020.

China-linked platforms also shared unconfirmed information or academic findings that had not been peer-reviewed if they supported China's talking points. For example (and as RT also did), Chinese media outlets posted reports about a study by a Stanford University professor who found that COVID-19 cases in Santa Clara County, California, were higher than the official count, thereby making the virus mortality rate appear lower than previously thought.[5] The outlet failed to report that the study had not been peer-reviewed at that time and that it evoked strong critique, with many in the scientific community questioning the study methods and findings.

Propaganda

Propaganda also constituted a large portion of China-linked information efforts. However, the content of propaganda messaging was not very diverse: Most of the messages focused on elevating China's global image, either through defending and glorifying China's actions in the pandemic or through strong criticism of the U.S. response (in explicit or implicit comparison with China). Contrasting China's superior pandemic response to "inadequate" U.S. countermeasures constituted a bulk of Chinese messaging within and about the United States. In these efforts, Chinese media often focused on the supposed failures of government officials in the United States, including the President, the Vice President, and the Secretary of State. At the same time, the Chinese information sphere reacted vehemently and in unison against any attacks that the United States made on China. Most often, such reactions involved some sort of mockery of U.S. attempts to discredit China and stories of successful Chinese counterpandemic efforts.[6]

---

[5] "U.S. COVID-19 Cases Are Higher Than Confirmed: Stanford Professor," CGTN America, via YouTube, May 5, 2020.

[6] For specific examples, see Amber Frankland, Bret Schafer, and Etienne Soula, "Hamilton Weekly Report, May 16–May 22, 2020," Alliance for Securing Democracy, May 26, 2020.

Trends in Content over Time

We noted a few patterns in China's messaging efforts over time (see Figure 3.1). Our analysis indicated that the priority of protection and defense of China's image remained clear and unwavering throughout material distributed between January 2020 and July 2020. However, Chinese messaging evolved from defense, explanation, and justification of Chinese actions to contrasting China's supposedly successful counter-COVID-19 efforts with the supposedly unsuccessful efforts of the United States. Later, this evolved into direct attacks and denigration of the U.S. pandemic response. By the end of the time frame that we analyzed, apparent Chinese messaging focused increasingly on painting a negative image of the United States without necessarily contrasting it with the positive developments in China.

As the global pandemic expanded, China-linked outlets focused on a wide variety of possibilities for the origins of the virus,[7] such as origination from an animal other than a bat and messages on the virus' artificial origins in and spread by the United States.[8] Over time, China-linked sources increasingly embraced the approach of denial of the virus' origins in China and polluting the information space with other seemingly plausible origin theories.

Instruments

COVID-19-related information efforts from China appeared to be at least somewhat coordinated across different media sources and accounts—from official state media outlets broadcasting in the United States to social media, whether through Chinese officials' Twitter accounts or trolls and bots. These various outlets exhibited informa-

---

[7]   For example, see Lisa Winter, "Chinese Officials Blame US Army for Coronavirus," *The Scientist*, March 13, 2020; Vanessa Molter and Renee DiResta, "Pandemics & Propaganda: How Chinese State Media Creates and Propagates CCP Coronavirus Narratives," *Misinformation Review*, June 8, 2020; David Gitter, Sandy Lu, and Brock Erdahl, "China Will do Anything to Deflect Coronavirus Blame," *Foreign Policy*, March 30, 2020; and Joshua Kurlantzick, "How China Ramped Up Disinformation Efforts During the Pandemic," Council on Foreign Relations, September 10, 2020.

[8]   Molter and DiResta, 2020.

Figure 3.1
2020 Trends in Content on COVID-19 from China and Its Agents

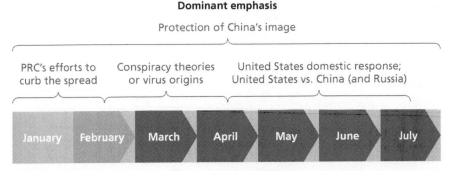

NOTE: Darker shades of blue indicate greater intensity of messaging at different points during the time frame that we analyzed. PRC = People's Republic of China.

tional unity, propagating similar messages and amplifying the official stance of the Chinese Communist Party (CCP). The main English-language state media that our analysis revealed were China Global Television Network (CGTN), China Central Television (CCTV), Xinhua News, and *Global Times* (see Table 3.2 for descriptions). In addition to their primary channels, these outlets actively utilized their Facebook and YouTube platforms to disseminate messages via direct sharing of content and paid political advertising.[9]

China's foreign policy officials played a significant role in propagating disinformation and propaganda. According to reports by the Alliance for Securing Democracy, the number of Chinese diplomatic Twitter accounts tripled between January 2020 and May 2020, and the Twitter output from China's embassies and ministries almost doubled in the same period.[10] A message disseminated via Twitter by a Chinese

---

[9] Molter and DiResta, 2020; Glen Tiffert, Renee DiResta, Carly Miller, Vanessa Molter, and John Pomfret, "Telling China's Story: The Chinese Communist Party's Campaign to Shape Global Narratives," Hoover Institution, July 21, 2020.

[10] Bret Schafer, "Beijing's Twitter Offensive in the Coronavirus 'Info War' on NBC News," Alliance for Securing Democracy, May 20, 2020; Anna Schecter, "China Launches New Twitter Accounts, 90,000 Tweets in COVID-19 Info War," NBC News, May 20, 2020.

Table 3.2
Description of the Principal China-Linked Disinformation Sources

Source	Description	Primary Audiences
CGTN	• International media organization • Claims to promote communication, understanding, and trust between China and the world	Left margins of U.S. political spectrum; administration skeptics; conspiracy seekers
Xinhua News	• China's official state-run press agency • Largest and most influential media organization in China and the largest news agency in the world in terms of number of news correspondents	
CCTV English	• Chinese public service broadcaster • Network of 50 channels and accessible to more than 1 billion viewers in six different languages	
Global Times	• Daily tabloid newspaper under the auspices of the CCP's *People's Daily* newspaper • Commentary on international issues from a Chinese nationalist perspective • Known for hawkish, provocative editorials	
China Daily	• CCP's English-language daily newspaper	

SOURCE: Global Engagement Center, 2020.

ambassador in one country was often picked up and shared by other Chinese officials around the world.[11] Such coordinated and unified messaging ensured that the information would spread widely and that Chinese messaging resonates in different parts of the world.

Trolls and bots were also activated in apparent coordination with the official sources of information. At different times between January 2020 and July 2020, multiple investigative sources (e.g., ProPublica, International Cyber Center, Bellingcat) identified fake and hacked Twitter accounts, Facebook pages, and YouTube channels that echoed Chinese messaging.[12] The origins of some of these Twitter accounts were later linked directly to China; the origins of others remain unknown. China-

---

[11] Mark Scott, "Chinese Diplomacy Ramps Up Social Media Offensive in COVID-19 Info War," *Politico*, April 29, 2020.

[12] See research by Jeff Kao and Mia Shuang Li, "How China Built a Twitter Propaganda Machine Then Let It Loose on Coronavirus," ProPublica, March 26, 2020; Elise Thomas and Albert Zhang, *COVID-19 Attracts Patriotic Troll Campaigns in Support of China's Geopolitical Interests*, Sydney, Australia: Australian Strategic Policy Institute, No. 2, April 23,

linked efforts also used messaging services, such as WhatsApp. For example, a China-linked source distributed messages about the impending nationwide lockdown in the United States through the WhatsApp mobile texting service.[13] Overall, despite the diversity of China-linked outlets and disinformation sources, on balance, the messages they produced were similar in content and in the audiences they targeted.

Audiences

Compared with Russia-linked messaging, China-linked messaging was more uniform across different outlets; this suggests that operators did not attempt to target specific polarized audiences or to purposefully appeal to a wide variety of audiences in the United States. However, during the time frame that we analyzed (January 2020 to July 2020), messages critical of the U.S. response to the pandemic might have resonated with critics of the Trump administration, those on the left of the U.S. political spectrum, and those concerned with the federal pandemic response. It is also possible that some of the messages about the origins of the virus could be attractive to conspiracy theory enthusiasts with different political views and affiliations. Overall, China-linked messaging could be of interest to U.S. audiences on the farther left of the political spectrum—Trump administration critics, conspiracy enthusiasts, and capitalism skeptics among them. See Figure 3.2 for a visual summary of U.S. audiences that could be receptive to the China-linked messaging on COVID-19. Unlike Russia-linked media sources (some of which enjoy popularity with a variety of U.S. audiences), it is not clear whether Chinese media summons enough attention among U.S. audiences to be considered influential. [14]

2020; and International Cyber Policy Centre, *COVID-19 Disinformation and Social Media Trends*, Australia: Australian Strategic Policy Institute, April 8–15, 2020.

[13] Wong, Rosenberg, and Barnes, 2020.

[14] Of note is that Chinese media efforts in the United States have predominantly targeted Chinese diasporas. However, the diaspora consumes media mostly in Chinese, so the targeting and influence on these populations remained beyond the scope of our analysis of the English-language sources only.

Figure 3.2
Audiences Potentially Vulnerable to Messages on COVID-19 from China and Its Agents and Proxies

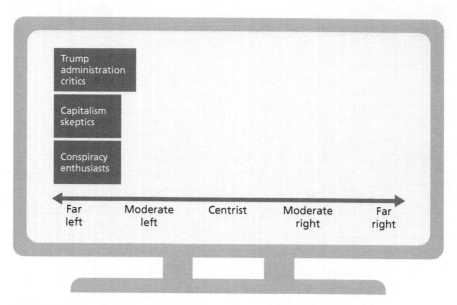

NOTES: The arrow symbolizes the U.S. political spectrum. The length of the blue shading of audience categories reflects where on the political spectrum the people within these categories could fall. For example, we expect that Trump administration skeptics among the moderate left likely would not be receptive to overtly China-linked messaging, but Trump administration skeptics farther to the left of the political spectrum might be. Note, also that U.S. audiences targeted by Chinese sources are limited to only three categories on the left side of the figure. Part of this might be because China exerted much of its effort attacking the U.S. administration in 2020 (i.e., Trump and Republicans).

Putting it Together: Objectives

Our analysis suggests that China pursued two principal strategic objectives: (1) to defend the image of China and its socioeconomic and political systems and (2) to diminish and denigrate the image of the United States and its socioeconomic and political systems. Our review indicated that the majority of Chinese messaging during the aforementioned time frame aimed to protect, defend, and enhance China's image (see Table 3.3 for examples)—which appears to have been

Table 3.3
Alignment Between Chinese Content and Likely Objectives

Messages	Likely Objectives
• It is getting harder to find the true origin of the novel coronavirus that has plagued the world and killed hundreds of thousands. • Genetic analysis indicates that the beginning of the pandemic dates to early October 2019—earlier than was previously thought.	Protect and bolster China's image
• What China does is much more effective than what the United States does in response to the pandemic. China should be a model for the United States. • The United States should be more organized; not having a centralized public health system is a problem. • The lack of centralized leadership in the United States makes it incapable of implementing multiple countermeasures. • Americans have been watching the world for two months and still doing nothing (in early March 2020). • The United States does not have effective ways to quarantine people and to monitor cases.	
• "Looking back at the timeline, anyone could say, 'If China did this at this time, things could have been better.' But don't forget that China entered uncharted territory and started the battle on its own. Other countries had a choice. • The group of people who accused China of being responsible for the pandemic are the very same ones who played it down in the first place. • The past is history, but the United States can still seize the present to treat more people instead of sparing no effort to find a scapegoat." ("Is China to Blame for the COVID-19 Pandemic?" CGTN, April 21, 2020)	
• Greed is guiding vaccine development efforts in the United States. • China will share a vaccine with the rest of the world, but the United States will keep it for itself.	Attack U.S. image
• Late response to COVID-19, failing to provide personal protective equipment, failing to provide people with access to health care, and ongoing debate about wearing a mask are all indicators that the United States has failed to respond appropriately.	

China's principal objective for its westward-looking media channels.[15] The messages that appeared to serve this objective were (1) claims of China's effective response to the pandemic internally and globally and (2) theories on the origins and spread of the virus. By disseminating multiple unfounded theories of the virus' origins and spread, these messages polluted the information environment, creating an situation in which it would be increasingly difficult for audiences to differentiate truth from myth and easier to see China as a benevolent (or even unfairly victimized) actor.

Defending China's image was coupled with attacking the image of the United States. Although we highlight the latter as a separate objective, we recognize that it is related and maybe even subservient to the central impetus of protecting and enhancing China's image globally. China-linked messaging pitted the U.S. and Chinese socioeconomic and political frameworks against each other. In doing so, China-linked messaging presented the United States and its capitalist democratic structures as failing to meet the pandemic challenge and care for the people. With time, the emphasis on discrediting the U.S. approaches to the pandemic and the U.S. image more generally became increasingly clear. China's future malign and subversive information efforts will likely continue to pursue the objective of damaging the U.S. image as independent from the need or urgency of protecting China's reputation.

China-linked messaging might have contributed to amplifying internal U.S. divides by pointing out U.S. weaknesses. However, the narrow focus of the messaging and the limited U.S. audiences to which it could appeal suggested that driving internal polarization was not one of China's principal objectives.

---

[15] Terry Flew, "CGTN: China's Latest Attempt to Win Friends and Influence People," *Asia Dialogue*, May 1, 2017; Merriden Varral, *Behind the News: Inside China Global Television Network*, Sydney, Australia: Lowy Institute, January 2020.

China Summary

Our analysis indicated that a variety of China-linked media sources disseminated diverse COVID-19-related content that could be categorized as disinformation or propaganda. The overall uniformity of messaging across the varied landscape of China-linked information sources and its evolution over time suggests that the CCP continued to prioritize elevating China's image as its primary strategic goal during the pandemic (see Figure 3.3).

Figure 3.3
Summary of Findings on China-Linked COVID-19 Messaging

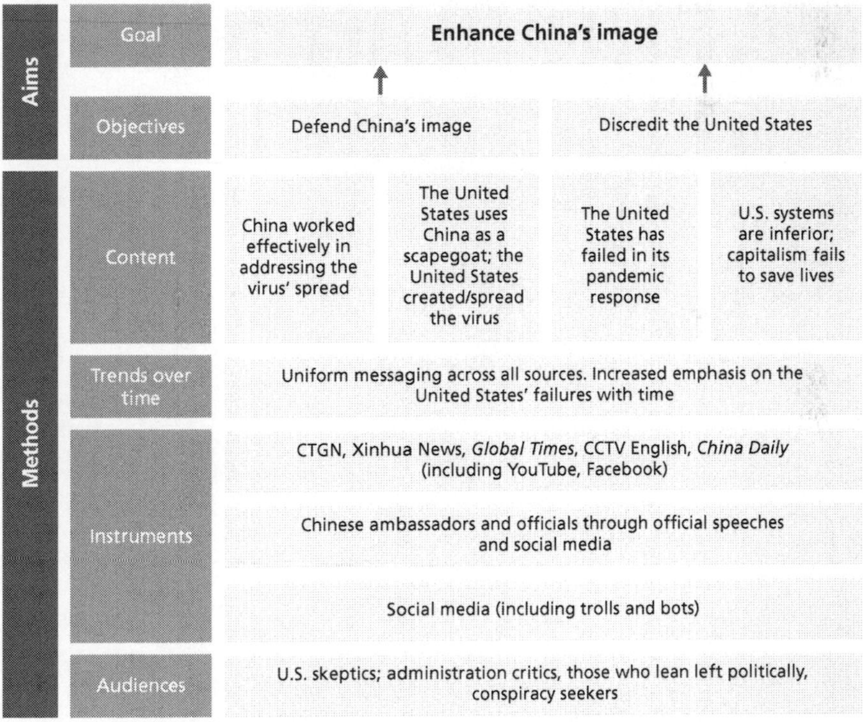

Aims	Goal	**Enhance China's image**			
		↑		↑	
	Objectives	Defend China's image		Discredit the United States	
Methods	Content	China worked effectively in addressing the virus' spread	The United States uses China as a scapegoat; the United States created/spread the virus	The United States has failed in its pandemic response	U.S. systems are inferior; capitalism fails to save lives
	Trends over time	Uniform messaging across all sources. Increased emphasis on the United States' failures with time			
	Instruments	CTGN, Xinhua News, *Global Times*, CCTV English, *China Daily* (including YouTube, Facebook)			
		Chinese ambassadors and officials through official speeches and social media			
		Social media (including trolls and bots)			
	Audiences	U.S. skeptics; administration critics, those who lean left politically, conspiracy seekers			

CHAPTER FOUR

Differences and Overlap in Russian and Chinese Messaging

Both Russia-linked and China-linked sources appear to have engaged in intensive malign and subversive information efforts targeting U.S. audiences and to have exhibited some similarities in their efforts. Both countries attempted to tarnish the reputation of the United States, most often by pointing out supposed inadequacies of the U.S. political and socioeconomic systems in response to the pandemic and by contrasting them with the supposedly more-effective centralized approaches of China and Russia. Russia-linked and China-linked sources also disseminated similar disinformation about the U.S. origins of the virus, the U.S. responsibility for the spread of the virus, and the malign work of U.S. biolabs around the world. It appears that the objective of denigrating the U.S. image and its socioeconomic and political systems is an area in which China and Russia found synergy in their COVID-19-related information efforts.

Both countries also appeared to promote China as an example of success in its fight to curb the coronavirus outbreak. Russia-linked sources likely shed a positive light on China's pandemic countermeasures more to disparage the image of the United States than to elevate China's image. Thus, it appears that Russia's focus on China was aimed at influencing the United States—seizing an opportunity to shed doubt on the effectiveness of the U.S. democratic and capitalist mechanisms compared with China's (and Russia's own) authoritarian ones.

There were, however, notable differences between Russian and Chinese efforts. Russia stayed clear of ideological and narrative consistency, deploying wide-ranging media and messages to target a variety of audiences across the U.S. political spectrum. As in several prior disinformation efforts,[1] Russia-linked messaging did not focus on promoting a certain point of view; rather it deployed a torrent of varied views simultaneously. Russia-linked sources gave voice to fringe concepts, airing opinions that did not appear on U.S. mainstream media. The Stanford Internet Observatory found that "almost 80% of opinion pieces about the United States appearing on RT's English-language channels were obviously negative in tone, and only 3% were positive."[2] Together with these findings, our analysis suggests that destabilizing the United States was the principal goal of Russian COVID-19-related malign and subversive information efforts targeting the United States during the time frame that we analyzed.

China-linked messaging, on the other hand, was unified across multiple information outlets (whether covert or overt), featured fewer narratives, and aligned closely with the ideologies and priorities of the CCP. China's apparent dominant objectives in malign and subversive information toward the United States were the defense of its own image and denigration of the U.S. image. Some of the China-promoted messages could be interpreted as pursuing objectives similar to Russia's (e.g., polluting the information space and widening U.S. internal divides). However, our analysis suggests that—for China—these apparent messaging aims were to defend China's image and/or disparaging the image of the United States. For Russia, discrediting the U.S. image, amplifying the internal divides, polluting information spaces, and sowing distrust all appeared to be independent, though complementary, objectives.[3]

---

[1]   For example, see Paul and Matthews, 2016.

[2]   Daniel Bush, "Virality Project (Russia): Penguins and Protests," Stanford Internet Observatory, blog post, June 9, 2020.

[3]   Using the message content, we analyzed what the apparent purpose of the message could be and the likely audiences with which such a message might have resonated. We then used these different types of information to analyze the possible objectives to which the Russian and Chinese malign information efforts were likely contributing. Finally, we compared the core content of the different messages prevalent among the entries for both countries and identified unique and overlapping themes. For more details, see the appendix.

The two countries also appeared to differ in the types of audiences that they targeted with their malign and subversive information efforts and in the level of sophistication with which they targeted these audiences. Russia-linked sources appear to have spread a wide and encompassing net to capture a variety of U.S. audiences on both the left and the right of the political spectrum. Furthermore, the links of the media sources to Russia are often hardly detectable. In contrast, China-linked media have openly focused on presenting and promoting China's point of view on global events and on enhancing China's image. Such emphasis limits China's likely audiences in the United States; more generally, the extent to which China's messaging efforts have been successful at garnering attention and influence among U.S. and Western audiences is unclear.[4] The apparently differing strategic goals of the two countries—and the sophistication of the media outlets they sponsor—might help explain the differences between the two countries' approaches.

Russia-linked and China-linked messaging also exhibited differences in how they evolved over time. China-linked COVID-19 messaging appeared to be consistent in its intensity throughout the time frame that we analyzed, from January 2020 to July 2020. On the other hand, the intensity of Russia-linked COVID-19-related messaging subsided starting in approximately May 2020 as the focus of messages shifted toward other (likely more divisive) themes linked to racial tensions in the United States. Furthermore, Russia-linked sources appeared to rely more heavily on conspiracy theories earlier in the time frame that we analyzed, and then shifted toward propaganda related to U.S. challenges in response to the virus.[5] China-linked sources followed a some-

---

[4]   Similar to China's approach to international messaging (see Flew, 2017; and Varral, 2020) Russia's past approach also focused on enhancing its image and promoting its point of view globally. When Russia realized that its self-promotion was not attracting Western audiences, it switched approaches and rebranded Russia Today into RT (I. Yablokov, "Conspiracy Theories as a Russian Public Diplomacy Tool: The Case of Russia Today (RT)," *Politics*, Vol. 35, No. 3–4, 2015).

[5]   "EEAS Special Report Update: Short Assessment of Narratives and Disinformation Around the COVID-19 Pandemic, Update: 23 April–18 May 2020," EUvsDISINFO, May 20, 2020.

what similar trend; focus on U.S. failures increased over time, but operators continued to disseminate conspiracy theories on the origins of the virus throughout the time frame that we analyzed.

Finally, there appears to be an important distinction between the apparent strategic goals of Russia and China. For Russia, the strategic goal of its malign and subversive information efforts appears to be the destabilization of the United States. China uses similar messaging, instruments, and approaches as Russia and its agents, but China's core strategic goal appears to be enhancing its own global standing. Figure 4.1 summarizes the similarities and differences between Russia's and China's messaging objectives.

Figure 4.1
Similarities and Differences Between Russia's and China's Messaging Objectives and Goals

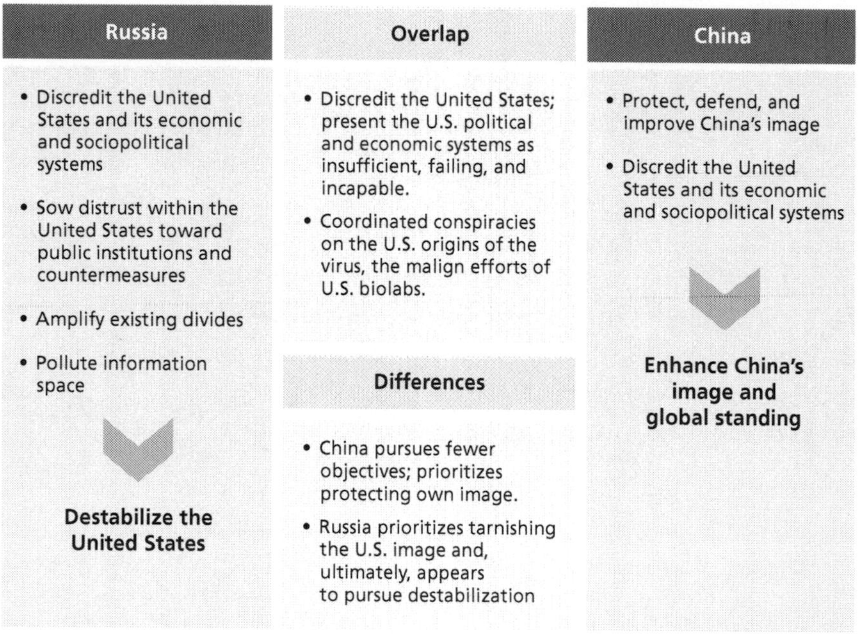

Russia	Overlap	China
• Discredit the United States and its economic and sociopolitical systems • Sow distrust within the United States toward public institutions and countermeasures • Amplify existing divides • Pollute information space	• Discredit the United States; present the U.S. political and economic systems as insufficient, failing, and incapable. • Coordinated conspiracies on the U.S. origins of the virus, the malign efforts of U.S. biolabs.	• Protect, defend, and improve China's image • Discredit the United States and its economic and sociopolitical systems
Destabilize the United States	**Differences** • China pursues fewer objectives; prioritizes protecting own image. • Russia prioritizes tarnishing the U.S. image and, ultimately, appears to pursue destabilization	**Enhance China's image and global standing**

CHAPTER FIVE

Discussion and Implications

Examining Russia's and China's COVID-19-related malign and subversive information efforts during a rapidly developing global crisis offers insight into how these two countries used this event to pursue their strategic objectives in the information space. Information on the tactics, tools, audiences, and trends involved during these campaigns can inform future countermessaging efforts.

Understanding and Countering Subversive Information Efforts

Countering apparent Russian and Chinese malign and subversive information efforts will require messaging campaigns that address the capabilities and thematic emphasis of each of these actors. Future research can inform these campaigns. Our research suggests that Russia and its agents appear to use an informational cacophony to promote Russia's goals and that China and its agents have, thus far, worked to build a uniform narrative. For example, our analysis addressing COVID-19 efforts suggests that Russia-linked sources consistently engaged in propaganda to emphasize negative developments within the United States and tarnish its image but also actively relied on disinformation to capitalize on the crisis and to amplify internal divides within the United States. In the case of China, disinformation and propaganda involving COVID-19 appeared to serve the goal of defending and enhancing the country's image. The distinctive uses of malign and subversive information efforts by the two countries could necessitate discrete approaches to counter each of them. Future research should explore the degree to which Russia and China continue to pursue distinctive uses of malign and subversive information efforts. The degree to which these uses are determined by each country's capabilities and evolving strategic goals should also be explored.

Russian Efforts

Russia and its agents continue to draw from its previous approach to malign and subversive information efforts. During the pandemic, Russia honed its existent approaches to disseminating malign and subversive information in support of its objectives. It will likely continue to build on and expand the same repertoire in the future. Specifically, Russia-linked sources are likely to continue to look for ways to destabilize the United States, taking advantage of public health and social issues that capture audience attention. Not bound to any ideology other than anti-Americanism in its efforts and with no apparent desire to make the West "like" it, Russia will likely remain relatively unconstrained in the variety, creativity, and boldness of its malign and subversive information efforts. Past work—based on observations that Russia continues to use similar tactics to pursue similar goals as those previously observed—proposed several avenues to addressing Russia's information efforts that can still be applied,[1] such as strategies for countering Russia's social media influence and existing tools for combating online disinformation.[2] For example, countermeasures could be

- forewarning audiences about the messages and categories of messages that Russia and its agents have and are likely to disseminate about COVID-19

---

[1]   Paul and Matthews, 2016.

[2]   Elizabeth Bodine-Baron, Todd C. Helmus, Andrew Radin, and Elina Treyger, *Countering Russian Social Media Influence*, Santa Monica, Calif.: RAND Corporation, RR-2740-RC, 2018; Jennifer Kavanagh, Samantha Cherney, Hilary Reininger, and Norah Griffin, *Fighting Disinformation Online: Building the Database of Web Tools*, Santa Monica, Calif.: RAND Corporation, RR-3000-WFHF, 2020.

- when possible, avoiding ways of directly refuting—and thereby repeating—false or misleading messages disseminated about COVID-19
- rapidly and repetitively disseminating correct information about COVID-19.

Chinese Efforts

China's malign and subversive information efforts involving U.S. audiences do not appear to be a substantial imminent threat to the United States. China's focus on enhancing its image and standing is not new, but multiple sources and our own analysis suggest that China has acted more aggressively during this crisis than in the past[3]—expanding both state platforms and covert information efforts through networks of trolls and bots. Our analysis also seems to indicate that, as the relationship between the United States and China becomes more strained, China's informational efforts targeting U.S. audiences could adopt new approaches and new goals (i.e., beyond efforts to defend and enhance its reputation)—possibly with U.S. destabilization among them. If this is the case, China might adopt Russia's playbook more fully and diversify its U.S.-facing messaging, tools, and targeted audiences.[4] At that point, similar avenues as those used for Russia could be used to counter China's new efforts. At present, however, China's malign and subversive information efforts involving COVID-19 appear unlikely to have a substantial impact on large proportions of U.S. audiences.

Counterefforts
Audience Consideration
Our analyses suggest that the COVID-19-related malign and subversive information efforts from Russia and China might have targeted different types of U.S. audiences: The variety of disinformation and propa-

---

[3] Kurlantzick, 2020; Kathy Gilsinan, " How China Is Planning to Win Back the World," *The Atlantic*, May 28, 2020; Clint Watts, "Triad of Disinformation: How Russia, Iran, & China Ally in a Messaging War Against America," Alliance for Securing Democracy, May 15, 2020.

[4] Kurlantzick, 2020; Jessica Brandt and Torrey Taussig, "The Kremlin's Disinformation Playbook Goes to Beijing," Brookings Institution, blog post, May 19, 2020.

ganda had the potential to influence relatively diverse U.S. audiences. Rather than one group serving as the primary target, multiple—often dissimilar—groups might be influenced by false and misleading messages from these countries. Documenting which audiences within the United States were and are most likely to be exposed to and influenced by these efforts is the sort of research that could help tailor countermeasures for specific groups.

Profiling Russian and Chinese Sources

Countermessaging efforts also would benefit from profiling of Russian and Chinese sources that are known to create and disseminate disinformation and propaganda. In addition to informing targeted audiences about known sources of disinformation and propaganda, profiling the malign and subversive information being disseminated by these sources could assist with developing better efforts to offset these messages. In other words, knowing where to look for these messages allows individuals and groups to find, analyze, and counter them.

Both countries disseminated messages through a wide variety of channels and platforms, including social media. Partially in response to public pressure and congressional oversight, social media platforms (such as Facebook and Twitter) have conducted a series of highly publicized takedowns of accounts linked to coordinated malign and subversive information campaigns, including networks of accounts linked to China and Russia.[5] However, both Russia and China appear to have directed many of their COVID-19 disinformation efforts to channels that are more traditional (and more difficult to block or take down), such as state-sponsored media websites and blogs. Our analysis focused on such channels and suggests that continuing to monitor and analyze these channels will be important in the days and years to come.

Automated monitoring and analysis could assist with early responses to Chinese and Russian malign and subversive informa-

---

[5] See Andrew Hutchinson, "Twitter and Facebook Announce Over 6,000 Account Removals Related to Political Manipulation," *Social Media Today*, December 21, 2019. Also see Arjun Kharpal, "Twitter Takes Down China-Linked Accounts Spreading Disinformation on Hong Kong and Coronavirus," CNBC, 2020; and Lisa Vaas, "Fake News Facebook Accounts Used Coronavirus to Attract Followers," *Naked Security*, May 7, 2020.

tion campaigns. Although a great deal of big-data analysis of disinformation has focused on social media platforms,[6] continued monitoring and analysis of all channels will be important in the production of early, accurate profiles of new disinformation campaigns. New capabilities for collecting and analyzing this information, such as computational linguistic analysis, have already shown promise in helping to characterize the rhetorical strategies used by malign actors to gain influence.[7] Such capabilities, when coupled with interactive user dashboards and provided with the capabilities to detect emerging trends and new topics in the information sphere, suggest promise as tools for countering the evolving campaigns of U.S. adversaries and malign actors.

Amplification of Messages

China and Russia appear to amplify one another's messages when such actions are advantageous to them. This could eventually lead to some collaboration, albeit limited in nature. The overlap in informational objectives during the pandemic crisis suggests that, if Russia's and China's goals remain complementary, the two countries might expand their collaboration in the information space in the future. Both countries might benefit from enhanced alliance in several ways: gaining a greater power to pollute the information space and drown out opposing world views; leveraging each other's platforms and technologies to create controlled and censored global information ecosystems; promoting each other's messaging to the audiences with which each country holds the greatest informational know-how (e.g., Russia in the United States and Europe; China in Asia, Latin America, and Africa); and

---

6   See Joshua A. Tucker, Andrew Guess, Pablo Barbera, Cristian Vaccari, Alexandra Siegel, Sergey Sanovich, Denis Stukal, and Brendan Nyhan, "Social Media, Political Polarization, and Political Disinformation: A Review of the Scientific Literature," *SSRN Electronic Journal*, March 19, 2018.

7   Elizabeth Bodine-Baron, Todd C. Helmus, Madeline Magnuson, and Zev Winkelman, *Examining ISIS Support and Opposition Networks on Twitter*, Santa Monica, Calif.: RAND Corporation, RR-1328-RC, 2016.

joining efforts in creating apps that would collect public information that can be leveraged later.[8]

At the same time, Russia and China will prioritize their own interests,[9] which could place limits on collaboration. Results from secondary analyses are consistent with our findings. For example, in their analysis of the future synergies between Russia and China, Dobbins and colleagues commented, "while their relationship is indeed growing across military, economic, and political dimensions, it is still anchored more in shared grievances than in common visions."[10] Russia's vision appears to be destruction of the U.S.-guided world order through destabilization of the United States specifically and the West more generally. By contrast, China's vision appears to be to grow its own global dominance and enhance its image through strengthening its ties to the West and changing the status quo—but not shattering the world order from which it has ultimately benefited.[11] Understanding these divergencies in Russia's and China's efforts should guide the U.S. response and action in the information space.

Public Health Messaging

Public health messaging should account for potential impacts of Russian and Chinese messaging on vaccination uptake. The effects of Chinese and Russian malign and subversive information campaigns could have implications for U.S. public health outcomes. U.S. vacci-

---

[8] See Daniel Kliman, Andrea Kendall-Taylor, Kristine Lee, Joshua Fitt, and Carisa Nietsche, *Dangerous Synergies: Countering Chinese and Russian Digital Influence Operations*, Washington, D.C.: Center for a New American Security, May 7, 2020.

[9] Andrea Kendall-Taylor and David Shullman, "How Russia and China Undermine Democracy: Can the West Counter the Threat?" *Foreign Affairs*, October 2, 2018.

[10] See James Dobbins, Howard J. Shatz, and Ali Wyne, "China and Russia Pose Different Problems for the US. They Need Different Solutions," *The Diplomat*, April 18, 2019b. For more details, see James Dobbins, Howard J. Shatz, and Ali Wyne, *Russia Is a Rogue, Not a Peer; China Is a Peer, Not a Rogue: Different Challenges, Different Responses*, Santa Monica, Calif.: RAND Corporation, PE-310-A, 2019a.

[11] Dan Coats, Office of the Director of National Intelligence, "Annual Threat Assessment, Opening Statement," testimony before the U.S. Senate Committee on National Intelligence Washington, D.C., on January 29, 2019. For a detailed discussion of broader implications of the differences in Russia's and China's world visions, see Dobbins et al., 2019a.

nation hesitancy for COVID-19 has been as high as 40 percent,[12] and decisions regarding mask-wearing and business shutdowns have been highly polarizing.[13] Although these phenomena are likely rooted in domestic disinformation and the prolonged lack of consistent public health messaging, these divides might make U.S. audiences particularly vulnerable to relevant disinformation stemming from Russia and China. Future research that examines how and whether foreign disinformation has contributed to public health behaviors could help tailor ongoing public health messaging about COVID-19. For example, targeted messaging campaigns from Russia and China might have instilled specific false beliefs about COVID-19, distrust in the U.S. government, and politically polarized reactions to the pandemic; all of these efforts might need to be countered in different ways for different demographic groups. Notably, prior research suggests that public health messaging often does not benefit from directly challenging false beliefs but instead from building on best practices in health communication, such as expressing empathy and fostering dialogue rather than pursuing more-confrontational "myth-busting" strategies or approaches that might risk labeling certain demographic groups as falling victim to foreign influence.[14]

Caveats and Next Steps

Previous research has also proposed that China and Russia have different strategic approaches to messaging (i.e., preserving the world order and ascending the ranks versus disrupting the world order), at least for

---

[12] Philip Elliott, "Science Delivers the COVID-19 Vaccine. Too Bad Not Enough People Want It," *Time*, December 4, 2020.

[13] "Republicans, Democrats Move Even Further Apart in Coronavirus Concerns," Pew Research Center, June 25, 2020.

[14] Nour Mheidly, and Jawad Fares, "Leveraging Media And Health Communication Strategies to Overcome the COVID-19 INFODEMIC," *Journal of Public Health Policy*, Vol. 41, No. 4, December 1, 2020.

the time being.[15] In our analysis, we identified two trends that have not been documented elsewhere (to our knowledge):

- Chinese and Russian messaging evolved in May 2020 from conspiracy theories to critiques of U.S. responses to COVID-19.
- Russian messaging had a much wider variation than Chinese messaging, reaching highly varied U.S. audiences.

It will be important to test these observations with follow-on quantitative analysis of news, blogs, and other sources. Such organizations as Graphika have proven the utility of using network-mapping on Twitter data to examine COVID-19 disinformation transmission over time.[16] Along with network analysis of social media data, we will also leverage computational linguistics to further explore the rhetorical and topical moves that Chinese and Russian efforts continue to employ in the COVID-19 "infodemic." This will include consideration of the messaging about the United States promoted by Russia- and China-linked sources, and we will share our results in a future report.

---

[15] Others have pointed out that China could take a much more aggressive and disruptive stance in the next few years, given that its window of opportunity for global dominance could be shrinking. For example, see Michael Beckley and Hal Brands, "Competition with China Could Be Short and Sharp: The Risk of War Is Greatest in the Next Decade," *Foreign Affairs*, December 17, 2020.

[16] Melanie Smith, Erin McAweeney, and Lea Ronzaud, "The COVID-19 'Infodemic': A Preliminary Analysis of the Online Conversation Surrounding the Coronavirus Pandemic," *Graphika*, 2020.

Methods

To examine apparent Russian and Chinese malign and subversive information dissemination to U.S. audiences on COVID-19, we used several strategies. First, we searched existing databases that systematically documented Russian or Chinese efforts. We relied most on English-language information from the following three sources: EUvsDisinfo, the Empirical Studies of Conflict (ESOC) COVID-19 Disinformation catalog, and Hamilton 2.0 dashboard.[1] EUvsDisinfo was established in 2015 by a task force of the European External Action Service that was set up to address Russian disinformation campaigns, and the EUvsDisinfo database of Russian disinformation—identifying dates, titles, outlets, and countries of messages—is updated weekly. A subset of this overall database addresses COVID-19 disinformation. ESOC began in 2009 and compiles micro-level data for research and policy purposes. The ESOC COVID-19 catalog was collected in collaboration with Microsoft Research. The Hamilton 2.0 dashboard is a project within the German Marshall Fund of the United States that provides summaries of narratives promoted by Russia, China, and Iran.

The EUvsDisinfo database focused only on messaging connected to Russia. By contrast, the ESOC catalog and the Hamilton 2.0 dashboard covered messaging from both Russia-linked sources and China-linked ones. The EUvsDisinfo database and ESOC catalog categorized messages as false (when appropriate and possible). The Hamilton 2.0 dashboard provided all highly viewed messages from Russian or Chi-

---

[1] "Coronavirus: Stay up to Date," undated; Shapiro, Oledan, and Siwakoti, 2020; Alliance for Securing Democracy, undated.

nese sources. When labeling information, both the EUvsDisinfo database and the ESOC COVID-19 catalog provided explanations for why such information was considered inaccurate or malign. These justifications included citing sources that provided evidence or rationale for why the information promoted by a Russia-linked or China-linked source was inaccurate. Furthermore, when attributing such false information to Russia or China, both databases provided links to the sources that promoted these messages.

We also searched English-language media channels linked to Russia and China using Boolean search terms, such as "coronavirus AND United States" or "COVID AND United States." We sought two types of sources: those formally linked to Russia and China (e.g., RT and CGTN, respectively) and those with indirect links to the Russian or Chinese government or Russian or Chinese networks (e.g., SouthFront or Global Research, respectively).[2] After identifying COVID-19-related content on these channels, we then used common fact-checking websites, such as factcheck.org and snopes.com, and additional web searches to identify whether information was factually accurate.

We made some assumptions about whether U.S. audiences could be exposed to the messages from Russian or Chinese sources we considered. In some cases, such as Russia's RT America or China's CGTN America, the United States is clearly identified as the target audience. In other cases, such as Russia's Sputnik or China's Xinhua News, we assumed U.S. audiences might consume information propagated by these media, which are accessible within the United States and deliver their content in English. Similarly, we considered English-language tweets from Chinese foreign officials around the world as information that could reach U.S. audiences.

We also reviewed secondary sources, such as scholarly analyses, government documents, and research reports to gather additional information about the COVID-19-related messages promoted to U.S. audiences by Russia, China, and their agents. In doing so, we searched

---

[2] Links to either Russia or China were established by third parties, such as EUvsDisinfo and the U.S. Department of State.

for themes that appeared in various independent English-language analytical accounts and for accounts that show clear links to the Russian or Chinese sources.

Finally, we also reviewed existing analyses on the instruments (e.g., media outlets, trolls and bots) that Russia, China, and their agents employed in spreading malign or subversive information to U.S. audiences in the first six months of the coronavirus crisis. To help calibrate this search, we referred to a variety of sources, such as Google searches of the materials published by independent think tanks, research institutes, and investigative journalists; searches for academic work using Google Scholar; and engagement with relevant webinars and panel discussions.

To derive the possible intent behind and the target audiences for the variety of messages promoted by the Russia-linked and China-linked outlets, we used the following two questions: *What is the purpose of promoting this messaging? What effect might this have on audiences?* Our analysis was grounded in previous analytical work on the intent behind Russian messaging in Europe and the hostile information measures both by China and Russia.[3]

In addition, insights from psychological science also guided our analyses of possible audiences for (and psychological effects of) Russia-linked and China-linked messaging. For example, research on confirmation bias suggests that in seeking information, individuals are motivated to avoid cognitive dissonance—i.e., confronting the reality that one's convictions or actions might be wrong—and to pay more attention to information that aligns with their world views.[4] People are particularly motivated to seek attitude-consistent information when there is an abundance of such messaging.[5] Psychological research fur-

[3] For Russian messaging, see Matthews et al., forthcoming; for hostile information measures, see Mazarr et al., 2019.

[4] L. Festinger, *A Theory of Cognitive Dissonance*, Stanford, Calif.: Stanford University Press, 1957; Silvia Knobloch-Westerwick, Cornelia Mothes, and Nick Polavin, "Confirmation Bias, Ingroup Bias, and Negativity Bias in Selective Exposure to Political Information," *Communication Research*, Vol. 47, No. 1, 2020.

[5] Knobloch-Westerwick, Mothes, and Polavin, 2020.

ther suggests that negative news garners more attraction and preference than does positive news.[6] Consistent and isolated exposure to attitude-congruent information (i.e., consuming information in echo chambers) likely leads to further polarization.[7]

To facilitate the analysis, all findings were organized within two country-specific spreadsheets: one for Russia and one for China.[8] We identified the date and core content of each entry, categorized it into a disinformation or propaganda effort, listed its source, and noted whether the information in each specific message was fact-checked. Using the message content, we analyzed what the apparent purpose of the message could be and the likely audiences with which such a message might have resonated. We then used these different types of information to analyze the possible objectives to which the Russian and Chinese malign information efforts were likely contributing. Finally, we compared the core content of the different messages prevalent among the entries for both countries and identified unique and overlapping themes.

---

[6]   Dolf Zillmann, Lei Chen, Silvia Knobloch, and Coy Callison, "Effects of Lead Framing on Selective Exposure to Internet News Reports," *Communication Research*, Vol. 31, No. 1, February 2004; Gunther Lengauer, Frank Esser, and Rosa Berganza, "Negativity in Political News: A Review of Concepts, Operationalizations and Key Findings," *Journalism*, Vol. 13, No. 2, February 2012.

[7]   Fryer, Jr., Harms, and Jackson, 2019.

[8]   We listed 80 entries in the spreadsheet addressing Russia-connected efforts, and we provide 37 entries in the spreadsheet for China-connected efforts. Notably, we focused on identifying and analyzing the content of messages, not the overall density or spread of messages. Therefore, we did not perform additional analyses, such as a density timeline.

References

Alliance for Securing Democracy, "Hamilton 2.0 Dashboard," German Marshall Fund of the United States, undated. As of November 20, 2020:
https://securingdemocracy.gmfus.org/hamilton-dashboard/

"America's China Blame Game Masks COVID-19 Failures (full show)," RT America, via YouTube, May 1, 2020. As of January 5, 2021:
https://www.youtube.com/watch?v=nBVF-2IMrZw

Amin, Avnika B., Robert A. Bednarczyk, Cara E. Ray, Kala J. Melchiori, Jesse Graham, Jeffrey R. Huntsinger, and Saad B. Omer, "Association of Moral Values with Vaccine Hesitancy," *Nature Human Behaviour*, Vol. 1, No. 12, December 2017, pp. 873–880.

Andrews, Peter, "COVID-19's Meant to Be a New Black Death, but in Britain No More People Are Dying Than NORMAL. What Does This Say About the Virus?" RT, March 31, 2020. As of January 5, 2021:
https://www.rt.com/op-ed/484548-coronavirus--people-die-outcome/

Armstrong, Martin, and Joaquin Flores, "World Health Organization Demanded Sweden Lock Down to Cover Its Own Fraud," Fort Russ News, April 17, 2020. As of January 5, 2021:
https://fort-russ.com/2020/04/
world-health-organization-demanded-sweden-lock-down-to-cover-its-own-fraud/

Baumgaertner, Bert, Julie E. Carlisle, and Florian Justwan, "The Influence of Political Ideology and Trust on Willingness to Vaccinate," *PloS One*, Vol. 13, No. 1, January 25, 2018.

Beckley, Michael, and Hal Brands, "Competition with China Could Be Short and Sharp: The Risk of War Is Greatest in the Next Decade," *Foreign Affairs*, December 17, 2020. As of December 18, 2020:
https://www.foreignaffairs.com/articles/united-states/2020-12-17/
competition-china-could-be-short-and-sharp

Bessi, Alessandro, Fabiana Zollo, Michela Del Vicario, Michelangelo Puliga, Antonio Scala, Guido Caldarelli, Brian Uzzi, and Walter Quattrociocchi, "Users Polarization on Facebook and YouTube," *PloS One*, Vol. 11, No. 8, August 23, 2016.

Bodine-Baron, Elizabeth, Todd C. Helmus, Madeline Magnuson, and Zev Winkelman, *Examining ISIS Support and Opposition Networks on Twitter*, Santa Monica, Calif.: RAND Corporation, RR-1328-RC, 2016. As of March 10, 2021: https://www.rand.org/pubs/research_reports/RR1328.html

Bodine-Baron, Elizabeth, Todd C. Helmus, Andrew Radin, and Elina Treyger, *Countering Russian Social Media Influence*, Santa Monica, Calif.: RAND Corporation, RR-2740-RC, 2018. As of March 10, 2021: https://www.rand.org/pubs/research_reports/RR2740.html

Brandt, Jessica, and Torrey Taussig, "The Kremlin's Disinformation Playbook Goes to Beijing," Brookings Institution, blog post, May 19, 2020. As of November 20, 2020: https://www.brookings.edu/blog/order-from-chaos/2020/05/19/the-kremlins-disinformation-playbook-goes-to-beijing/

Bruder, Martin, Peter Haffke, Nick Neave, Nina Nouripanah, and Roland Imhoff, "Measuring Individual Differences in Generic Beliefs in Conspiracy Theories Across Cultures: Conspiracy Mentality Questionnaire," *Frontiers in Psychology*, Vol. 4, No. 225, April 30, 2013.

Bush, Daniel, "Virality Project (Russia): Penguins and Protests," Stanford Internet Observatory, blog post, June 9, 2020. As of November 20, 2020: https://cyber.fsi.stanford.edu/io/news/penguins-and-protests-rt-and-coronavirus-pandemic

Butler, Kiera, "The Anti-Vax Movement's Radical Shift from Crunchy Granola Purists to Far-Right Crusaders," *Mother Jones*, June 18, 2020. As of November 19, 2020: https://www.motherjones.com/politics/2020/06/the-anti-vax-movements-radical-shift-from-crunchy-granola-purists-to-far-right-crusaders/

Butler, Phil, "Is COVID-19 Devouring What's Left of the Trump Presidency?" *New Eastern Outlook*, March 16, 2020. As of January 5, 2021: https://journal-neo.org/2020/03/16/is-covid-19-devouring-what-s-left-of-the-trump-presidency/

"China Delayed Releasing Coronavirus Information for Days at Start of Outbreak, Frustrating the WHO," *Los Angeles Times*, June 2, 2020.

"Chinese Official: World Needs to Know When Coronavirus Started in U.S.," CGTN, April 29, 2020. As of November 19, 2020: https://news.cgtn.com/news/2020-04-29/Chinese-official-World-needs-to-know-when-coronavirus-started-in-U-S--Q5icZe5DeE/index.html

Coats, Dan, Office of the Director of National Intelligence, "Annual Threat Assessment, Opening Statement," testimony before the U.S. Senate Committee on National Intelligence, Washington, D.C., January 29, 2019. As of November 20, 2020:
https://www.dni.gov/files/documents/Newsroom/Testimonies/ 2019-01-29-ATA-Opening-Statement_Final.pdf

"Commentary: A Disease-Ridden U.S. Fails World in Anti-Virus Cooperation," Xinhua, November 15, 2020. As of January 12, 2021:
http://www.xinhuanet.com/english/2020-11/15/c_139517314.htm

"The Community of Collapse," EUvsDiSiNFO, March 27, 2020. As of November 19, 2020:
https://euvsdisinfo.eu/the-community-of-collapse/

"Coronavirus: Stay up to Date," EUvsDiSiNFO database, undated. As of November 20, 2020:
https://euvsdisinfo.eu/disinformation-cases/?text=coronavirus&date=

"COVID-19 Eroding Civil Liberties—Gone for Good?" RT America, via YouTube, March 31, 2020. As of January 5, 2021:
https://www.youtube.com/watch?v=TUl7kswYZBU

"COVID-19 Spread Started as Early as October 6, 2019: Study," CGTN, May 6, 2020. As of January 5, 2021:
https://news.cgtn.com/news/2020-05-06/Study-Coronavirus-quickly-spread-around-the-world-starting-late-2019-QgPUDNqhl6/index.html

Dimock, Michael, and Richard Wike, "America Is Exceptional in the Nature of Its Political Divide," *FactTank News*, Pew Research Center, November 13, 2020. As of November 19, 2020:
https://www.pewresearch.org/fact-tank/2020/11/13/america-is-exceptional-in-the-nature-of-its-political-divide/

Dobbins, James, Howard J. Shatz, and Ali Wyne, *Russia Is a Rogue, Not a Peer; China Is a Peer, Not a Rogue: Different Challenges, Different Responses*, Santa Monica, Calif.: RAND Corporation, PE-310-A, 2019a. As of November 10, 2020:
https://www.rand.org/pubs/perspectives/PE310.html

Dobbins, James, Howard J. Shatz, and Ali Wyne, "China and Russia Pose Different Problems for the US. They Need Different Solutions," *The Diplomat*, April 18, 2019b. As of November 20, 2020:
https://thediplomat.com/2019/04/ china-and-russia-pose-different-problems-for-the-us-they-need-different-solutions/

Douglas, Karen M., Joseph E. Uscinski, Robbie M. Sutton, Aleksandra Cichocka, Turkay Nefes, Chee Siang Ang, and Farzin Deravi, "Understanding Conspiracy Theories," *Political Psychology*, Vol. 40, No. 1, 2019, pp. 3–35.

"EEAS Special Report Update: Short Assessment of Narratives and Disinformation Around the COVID-19 Pandemic, Update: 23 April–18 May 2020," EUvsDISINFO, May 20, 2020. As of November 20, 2020: https://euvsdisinfo.eu/eeas-special-report-update-short-assessment-of-narratives-and-disinformation-around-the-covid19-pandemic-updated-23-april-18-may

Einstein, Katherine L., and David M. Glick, *Scandals, Conspiracies and the Vicious Cycle of Cynicism*, Chicago, Ill.: 2013 Annual Meeting of the American Political Science Association, working paper, August 29–September 1, 2013.

Elliott, Philip, "Science Delivers the COVID-19 Vaccine. Too Bad Not Enough People Want It," *Time*, December 4, 2020. As of December 18, 2020: https://time.com/5918040/coronavirus-vaccine-hesitancy/

Enders, Adam M., and Steven M. Smallpage, "Informational Cues, Partisan-Motivated Reasoning, and the Manipulation of Conspiracy Beliefs," *Political Communication*, Vol. 36, No. 1, 2019, pp. 83–102.

Festinger, Leo, *A Theory of Cognitive Dissonance*, Stanford, Calif.: Stanford University Press, 1957.

Flew, Terry, "CGTN: China's Latest Attempt to Win Friends and Influence People," *Asian Dialogue*, May 1, 2017. As of November 20, 2020: https://theasiadialogue.com/2017/05/01/cgtn-chinas-latest-attempt-to-win-friends-and-influence-people/

Flores, Joaquin, Tom Winter, and Drago Bosnic, eds., " About FRN," Fort Russ News, 2014. As of November 19, 2020: https://fort-russ.com/abou/

Frankland, Amber, Bret Schafer, and Matt Schrader, "Hamilton Weekly Report: May 2–8, 2020," Alliance for Securing Democracy, May 11, 2020. As of November 19, 2020: https://securingdemocracy.gmfus.org/hamilton-weekly-report-may-2-8-2020/

Frankland, Amber, Bret Schafer, and Etienne Soula, "Hamilton Weekly Report, May 16–May 22, 2020," Alliance for Securing Democracy, May 26, 2020. As of November 19, 2020: https://securingdemocracy.gmfus.org/hamilton-weekly-report-may-16-22-2020/

Fryer, Roland G., Jr., Philipp Harms, and Matthew O. Jackson, "Updating Beliefs When Evidence Is Open to Interpretation: Implications for Bias and Polarization," *Journal of the European Economic Association*, Vol. 17, No. 5, October 2019, pp. 1470–1501. As of November 20, 2020: https://www.doi.org/10.1093/jeea/jvy025

Galliford, Natasha, and Adrian Furnham, "Individual Difference Factors and Beliefs in Medical and Political Conspiracy Theories," *Scandinavian Journal of Psychology*, Vol. 58, No. 5, 2017, pp. 422–428.

Gilsinan, Kathy, " How China Is Planning to Win Back the World," *The Atlantic*, May 28, 2020. As of November 20, 2020:
https://www.theatlantic.com/politics/archive/2020/05/china-disinformation-propaganda-united-states-xi-jinping/612085/

Gitter, David, Sandy Lu, and Brock Erdahl, "China Will Do Anything to Deflect Coronavirus Blame," Foreign Policy, March 30, 2020. As of November 19, 2020:
https://foreignpolicy.com/2020/03/30/beijing-coronavirus-response-see-what-sticks-propaganda-blame-ccp-xi-jinping/

Global Engagement Center, *GEC Special Report: Pillars of Russia's Disinformation and Propaganda Ecosystem*, Washington, D.C.: U.S. Department of State, August 2020.

"GLOBALink | Brazilian Expert Applauds China-Brazil Cooperation in COVID-19 Vaccines," Xinhua, January 9, 2021. As of January 12, 2021:
http://www.xinhuanet.com/english/2021-01/09/c_139654345.htm

Golec de Zavala, Agnieszka, and Christopher M. Federico, "Collective Narcissism and the Growth of Conspiracy Thinking over the Course of the 2016 United States Presidential Election: A Longitudinal Analysis," *European Journal of Social Psychology*, Vol. 48, No. 7, 2018, pp. 1011–1018.

Goreis, Andreas, and Martin Voracek, "A Systematic Review and Meta-Analysis of Psychological Research on Conspiracy Beliefs: Field Characteristics, Measurement Instruments, and Associations with Personality Traits," *Frontiers in Psychology*, Vol. 10, No. 205, February 11, 2019.

Grzesiak-Feldman, Monika, and Monika Irzycka, "Right-Wing Authoritarianism and Conspiracy Thinking in a Polish Sample," *Psychological Reports*, Vol. 105, No. 2, 2009, pp. 389–393.

"How the US Overlooked the COVID Outbreak (RT documentary)," RT America, via YouTube, April 18, 2020. As of January 5, 2021:
https://www.youtube.com/watch?v=d-cQ5uVeiwo

"How the U.S. Response to COVID-19 Failed and Caused Thousands of Deaths," CGTN America, via YouTube, May 25, 2020. As of January 5, 2021:
https://www.youtube.com/watch?v=_Geb5l6Ymhw

Hu Xijin, "[#HuSays] Trump refused to participate in the #WHA and sent US HHS Secretary Alex Azar in his place . . . ," *Global Times* post on Facebook, May 20, 2020. As of January 12, 2021:
https://www.facebook.com/ads/library/?id=170182921102361

———, "The #US government's arrogance is leading the country down a road devoid of science or rationality . . . ," *Global Times* post on Facebook, August 5, 2020. As of January 12, 2021:
https://www.facebook.com/ads/library/?id=958699394604936

Hutchinson, Andrew, "Twitter and Facebook Announce Over 6,000 Account Removals Related to Political Manipulation," *Social Media Today*, December 21, 2019. As of August 15, 2020:
https://www.socialmediatoday.com/news/
twitter-and-facebook-announce-over-6000-account-removals-related-to-politi/
569560

Imhoff, Roland, and Martin Bruder, "Speaking (Un-) Truth to Power: Conspiracy Mentality as a Generalised Political Attitude," *European Journal of Personality*, Vol. 28, No. 1, 2014, pp. 25–43.

International Cyber Policy Centre, *COVID-19 Disinformation and Social Media Trends*, Barton, Australia: Australian Strategic Policy Institute, April 8–15, 2020.

"Is China to Blame for the COVID-19 Pandemic?" CGTN, April 21, 2020. As of March 20, 2021:
https://news.cgtn.com/news/2020-04-21/
Is-China-to-blame-for-the-COVID-19-pandemic--PRImh7pvMs/index.html

"Is There a Doctor in the House? Nevada State Medical Team That Banned Malaria Drug for Coronavirus Lacks Qualification," RT, March 25, 2020. As of January 5, 2021:
https://www.rt.com/usa/484111-nevada-doctors-unlicensed-chloroquine-ban/

Jackson, Dean, "Distinguishing Disinformation from Propaganda, Misinformation, and 'Fake News,'" Washington, D.C.: National Endowment for Democracy, Issue Brief, October 17, 2017. As of November 19, 2020:
https://www.ned.org/issue-brief-distinguishing-disinformation-from-pro
paganda-misinformation-and-fake-news/

Jost, John T., Jack Glaser, Arie W. Kruglanski, and Frank J. Sulloway, "Political Conservatism as Motivated Social Cognition," *Psychological Bulletin*, Vol. 129, No. 3, 2003, p. 339.

Kaiser, Jonas, and Adrian Rauchfleisch, "Unite the Right? How YouTube's Recommendation Algorithm Connects the US Far-Right," *Medium*, April 11, 2018.

Kao, Jeff, and Mia Shuang Li, "How China Built a Twitter Propaganda Machine Then Let It Loose on Coronavirus," *ProPublica*, March 26, 2020. As of November 20, 2020:
https://www.propublica.org/article/how-china-built-a-twitter-propaganda-
machine-then-let-it-loose-on-coronavirus

Kavanagh, Jennifer, Samantha Cherney, Hilary Reininger, and Norah Griffin, *Fighting Disinformation Online: Building the Database of Web Tools*, Santa Monica, Calif.: RAND Corporation, RR-3000-WFHF, 2020. As of March 10, 2021:
https://www.rand.org/pubs/research_reports/RR3000.html

Kavanagh, Jennifer, and Michael D. Rich, *Truth Decay: An Initial Exploration of the Diminishing Role of Facts and Analysis in American Public Life*, Santa Monica, Calif.: RAND Corporation, RR-2314-RC, 2018. As of March 15, 2021:
https://www.rand.org/pubs/research_reports/RR2314.html

Kendall-Taylor, Andrea, and David Shullman, "How Russia and China Undermine Democracy: Can the West Counter the Threat?" *Foreign Affairs*, October 2, 2018. As of November 20, 2020:
https://www.foreignaffairs.com/articles/china/2018-10-02/
how-russia-and-china-undermine-democracy?cid=int-lea&pgtype=hpg

Kharpal, Arjun, "Twitter Takes Down China-Linked Accounts Spreading Disinformation on Hong Kong and Coronavirus," CNBC, 2020. As of August 15, 2020:
https://www.cnbc.com/2020/06/12/twitter-takes-down-china-linked-accounts-spreading-disinformation.html

Kliman, Daniel, Andrea Kendall-Taylor, Kristine Lee, Joshua Fitt, and Carisa Nietsche, *Dangerous Synergies: Countering Chinese and Russian Digital Influence Operations*, Washington, D.C.: Center for a New American Security, May 7, 2020. As of November 20, 2020:
https://www.cnas.org/publications/reports/dangerous-synergies

Knobloch-Westerwick, Silvia, Cornelia Mothes, and Nick Polavin, "Confirmation Bias, Ingroup Bias, and Negativity Bias in Selective Exposure to Political Information," *Communication Research*, Vol. 47, No. 1, 2020, pp. 104–124.

Koenig, Peter, "The Coronavirus COVID-19 Pandemic: The Real Danger is Agenda ID2020," GlobalResearch, March 12, 2020. As of January 5, 2021:
https://www.globalresearch.ca/
coronavirus-causes-effects-real-danger-agenda-id2020/5706153

Korybko, Andrew, "The US' Medical Terrorism Against Syria Threatens to Make World War C Much Deadlier," One World Global Think Tank, May 20, 2020. As of January 5, 2021:
http://oneworld.press/?module=articles&action=view&id=1473

Kurlantzick, Joshua, "How China Ramped Up Disinformation Efforts During the Pandemic," *In Brief*, Council on Foreign Relations, September 10, 2020. As of November 20, 2020:
https://www.cfr.org/in-brief/
how-china-ramped-disinformation-efforts-during-pandemic

Laskin, Alexander V., "Defining Propaganda: A Psychoanalytic Perspective," *Communication and the Public*, Vol. 4, No. 4, 2019, pp. 305–314.

Lahrach, Y., and Adrian Furnham, "Are Modern Health Worries Associated with Medical Conspiracy Theories?" *Journal of Psychosomatic Research*, Vol. 99, 2017, pp. 89–94.

Le Yu, "Four Questions the US Must Answer Concerning COVID-19," *Global Times*, May 1, 2020. As of January 5, 2021:
https://www.globaltimes.cn/content/1187243.shtml

Lengauer, Gunther, Frank Esser, and Rosa Berganza, "Negativity in Political News: A Review of Concepts, Operationalizations and Key Findings," *Journalism*, Vol. 13, No. 2, February 2012, pp. 179–202.

Linvill, Darren L., and Patrick Warren, "Yes, Russia Spreads Coronavirus Lies. But They Were Made in America," *Washington Post*, April 2, 2020.

Matthews, Miriam, Alyssa Demus, Elina Treyger, Marek N. Posard, Hilary Reininger, and Christopher Paul, *Understanding and Defending Against Russia's Malign and Subversive Information Efforts in Europe*, Santa Monica, Calif.: RAND Corporation, RR-3160-EUCOM, forthcoming.

Mazarr, Michael J., Abigail Casey, Alyssa Demus, Scott W. Harold, Luke J. Matthews, Nathan Beauchamp-Mustafaga, and James Sladden, *Hostile Social Manipulation: Present Realities and Emerging Trends*, Santa Monica, Calif.: RAND Corporation, RR-2713-OSD, 2019. As of November 20, 2020:
https://www.rand.org/pubs/research_reports/RR2713.html

"Media Bias/Fact Check: The Most Comprehensive Media Bias Resource," *Sputnik News*, webpage, undated. As of November 19, 2020:
https://mediabiasfactcheck.com/sputnik-news/

Mheidly, Nour, and Jawad Fares, "Leveraging Media and Health Communication Strategies to Overcome the COVID-19 Infodemic," *Journal of Public Health Policy*, Vol. 41, No. 4, December 1, 2020, pp. 410–420.

Miller, Joanne M., Kyle L. Saunders, and Christina E. Farhart, "Conspiracy Endorsement as Motivated Reasoning: The Moderating Roles of Political Knowledge and Trust," *American Journal of Political Science*, Vol. 60, No. 4, 2016, pp. 824–844.

Molter, Vanessa, and Renee DiResta, "Pandemics & Propaganda: How Chinese State Media Creates and Propagates CCP Coronavirus Narratives," *Misinformation Review*, June 8, 2020. As of November 19, 2020:
https://misinforeview.hks.harvard.edu/article/pandemics-propaganda-how-chinese-state-media-creates-and-propagates-ccp-coronavirus-narratives/

OneWorld Global Think Tank, "About Us," webpage, undated. As of November 19, 2020:
https://oneworld.press/?module=tree&action=view&id=13

Paul, Christopher, and Miriam Matthews, *The Russian "Firehose of Falsehood" Propaganda Model: Why It Might Work and Options to Counter It*, Santa Monica, Calif.: RAND Corporation, PE-198-OSD, 2016. As of November 10, 2020:
https://www.rand.org/pubs/perspectives/PE198.html

Van Prooijen, Jan-Willem, and Michele Acker, "The Influence of Control on Belief in Conspiracy Theories: Conceptual and Applied Extensions," *Applied Cognitive Psychology*, Vol. 29, No. 5, 2015, pp. 753–761.

Van Prooijen, Jan-Willem, André P. M. Krouwel, and Thomas V. Pollet, "Political Extremism Predicts Belief in Conspiracy Theories," *Social Psychological and Personality Science*, Vol. 6, No. 5, 2015, pp. 570–578.

Propaganda, Misinformation, Disinformation & Fact Finding Resources, Detroit, Mich.: Wayne State, University Library System, Research Guides, 2020. As of November 19, 2020:
https://guides.lib.wayne.edu/c.php?g=401320&p=2729574

Qiao Collective [@qiaocollective], "China warns that it will not accept any US invasion of Venezuela," Twitter post, May 6, 2020. As of January 5, 2021:
https://twitter.com/qiaocollective/status/1258118753142943744

Rabinowitz, Mitchell, Lauren Latella, Chadly Stern, and John T. Jost, "Beliefs About Childhood Vaccination in the United States: Political Ideology, False Consensus, And the Illusion of Uniqueness," *PloS One*, Vol. 11, No. 7, July 8, 2016.

"Republicans, Democrats Move Even Further Apart in Coronavirus Concerns," Pew Research Center, June 25, 2020. As of December 5, 2020:
https://www.pewresearch.org/politics/2020/06/25/
republicans-democrats-move-even-further-apart-in-coronavirus-concerns/

Rettie, Hannah, and Jo Daniels, "Coping and Tolerance of Uncertainty: Predictors and Mediators of Mental Health During the COVID-19 Pandemic," *American Psychologist*, August 2020.

Richey, Sean, "A Birther and a Truther: The Influence of the Authoritarian Personality on Conspiracy Beliefs," *Politics & Policy*, Vol. 45, No. 3, 2017, pp. 465–485.

Rosenberg, Matthew, Nicole Perlroth, and David E. Sanger, "'Chaos Is the Point': Russian Hackers and Trolls Grow Stealthier in 2020," *New York Times*, January 10, 2020.

Ross, John, "Human Life Must Trump Economics in a Pandemic. THIS Is Why China Is Succeeding in War on COVID-19 and US Is on Path to Disaster," RT, March 30, 2020. As of January 5, 2021:
https://www.rt.com/op-ed/484517-china-us-disastrous-pandemic-response/

Schafer, Bret, "Beijing's Twitter Offensive in the Coronavirus 'Info War' on NBC News," Alliance for Securing Democracy, May 20, 2020. As of November 20, 2020:
https://securingdemocracy.gmfus.org/bret-schafer-on-beijings-twitter-offensive-in-the-coronavirus-info-war-on-nbc-news/

Schecter, Anna, "China Launches New Twitter Accounts, 90,000 Tweets in COVID-19 Info War," NBC News, May 20, 2020. As of November 20, 2020: https://www.nbcnews.com/news/world/ china-launches-new-twitter-accounts-90-000-tweets-covid-19-n1207991

Scott, Mark, "Chinese Diplomacy Ramps Up Social Media Offensive in COVID-19 Info War," *Politico*, April 29, 2020. As of November 20, 2020: https://www.politico.eu/article/china-disinformation-covid19-coronavirus/

Shapiro, Jacob N., Jan Oledan, and Samikshya Siwakoti, *ESOC COVID019 Disinformation Tracking Report*, Princeton, N.J.: Empirical Studies of Conflict, Princeton University, 2020.

Smith, Melanie, Erin McAweeney, and Lea Ronzaud, "The COVID-19 'Infodemic': A Preliminary Analysis of the Online Conversation Surrounding the Coronavirus Pandemic, *Graphika*, 2020. As of September 10, 2020: https://public-assets.graphika.com/reports/Graphika_Report_Covid19_Infodemic.pdf

Solajic, Slobodan, "The COVID-19 Plandemic Is an Experiment in Manipulating the World," One World Global Think Tank, June 16, 2020. As of January 5, 2021: https://oneworld.press/?module=articles&action=view&id=1521

Spokesperson发言人办公室 [@MFA_China], "The local people are deeply concerned about the function, purpose & safety of the biological labs in former Soviet Union countries which were established by the US. The US should explain and address the concerns." Twitter post, April 29, 2020. As of January 5, 2021: https://twitter.com/MFA_China/status/1255470813266038785

Thomas, Elise, and Albert Zhang, *COVID-19 Attracts Patriotic Troll Campaigns in Support of China's Geopolitical Interests*, Sydney, Australia: Australian Strategic Policy Institute, No. 2, April 23, 2020. As of November 20, 2020: https://www.aspi.org.au/report/covid-19-disinformation

Tiffert, Glenn, Renee DiResta, Carly Miller, Vanessa Molter, and John Pomfret, "Telling China's Story: The Chinese Communist Party's Campaign to Shape Global Narratives," Hoover Institution, July 21, 2020. As of November 19, 2020: https://www.hoover.org/research/ telling-chinas-story-chinese-communist-partys-campaign-shape-global-narratives

"Tracking Disinformation and Conflict," *ESOC*, webpage, 2020. As of November 20, 2020: https://esoc.princeton.edu/projects/tracking-disinformation-and-conflict

"Trump's Anti-Coronavirus Farce," CGTN America, via YouTube, May 26, 2020. As of January 5, 2021: https://www.youtube.com/watch?v=Vdmm4H7jeLA

Tucker, Joshua A., Andrew Guess, Pablo Barbera, Cristian Vaccari, Alexandra Siegel, Sergey Sanovich, Denis Stukal, and Brendan Nyhan, "Social Media, Political Polarization, and Political Disinformation: A Review of the Scientific Literature," *SSRN Electronic Journal*, March 19, 2018. As of October 3, 2020: https://ssrn.com/abstract=3144139

Tutui, Viorel, "Some Reflections Concerning the Problem of Defining Propaganda," *Argumentum*, Vol. 15, No. 2, 2017, pp. 110–125.

"U.S. COVID-19 Cases Are Higher Than Confirmed: Stanford Professor," CGTN America, via YouTube, May 5, 2020. As of November 19, 2020: https://www.youtube.com/watch?v=EjP_i9fdhAo

Uscinski, Joseph E., Adam M. Enders, Casey Klofstad, Michelle Seelig, John Funchion, Caleb Everett, Stefan Wuchty, Kamal Premaratne, and Manohar Murthi, "Why Do People Believe COVID-19 Conspiracy Theories?" *Misinformation Review*, Vol. 1, No. 3, April 28, 2020. As of November 20, 2020: https://misinforeview.hks.harvard.edu/article/why-do-people-believe-covid-19-conspiracy-theories/

Uscinski, Joseph E. , Casey Klofstad, and Matthew D. Atkinson, "What Drives Conspiratorial Beliefs? The Role of Informational Cues and Predispositions," *Political Research Quarterly*, Vol. 69, No. 1, 2016, pp. 57–71.

Uscinski, Joseph E., and Joseph M. Parent, *American Conspiracy Theories*, New York: Oxford University Press, 2014.

Vaas, Lisa, "Fake News Facebook Accounts Used Coronavirus to Attract Followers," *Naked Security*, May 7, 2020. As of September 1, 2020: https://nakedsecurity.sophos.com/2020/05/07/fake-news-facebook-accounts-used-coronavirus-to-attract-followers/

Varral, Merriden, *Behind the News: Inside China Global Television Network*, Sydney, Australia: Lowy Institute, January 2020. As of November 20, 2020: https://www.lowyinstitute.org/publications/behind-news-inside-china-global-television-network#_edn92

Watts, Clint, "Triad of Disinformation: How Russia, Iran, & China Ally in a Messaging War Against America," Alliance for Securing Democracy, May 15, 2020. As of November 20, 2020: https://securingdemocracy.gmfus.org/triad-of-disinformation-how-russia-iran-china-ally-in-a-messaging-war-against-america/

Winter, Lisa, "Chinese Officials Blame U.S. Army for Coronavirus," *The Scientist*, March 13, 2020. As of November 19, 2020: https://www.the-scientist.com/news-opinion/chinese-officials-blame-us-army-for-coronavirus-67267

Wong, Edward, Matthew Rosenberg, and Julian E. Barnes, "Chinese Agents Helped Spread Messages That Sowed Virus Panic in U.S., Officials Say," *New York Times*, April 22, 2020, updated January 5, 2021.

Yablokov, I., "Conspiracy Theories as a Russian Public Diplomacy Tool: The Case of Russia Today (RT)," *Politics*, Vol. 35, No. 3–4, 2015, pp. 301–315.

Zhao Lijian [@zlj517], "This article is very much important to each and every one of us. Please read and retweet it. COVID-19: Further Evidence that the Virus Originated in the US," Twitter post, March 12, 2020. As of January 5, 2021: https://twitter.com/zlj517/status/1238269193427906560?lang=en

Zillmann, Dolf, Lei Chen, Silvia Knobloch, and Coy Callison, "Effects of Lead Framing on Selective Exposure to Internet News Reports," *Communication Research*, Vol. 31, No. 1, February 2004, pp. 58–81.